living with an old

David Alderton & Derek Hall

Hubble & Hattie

Photographs by Marc Henrie,
with additional material from Derek Hall

Gentle Dog Care

Hubble & Hattie

For more than nineteen years, the folk at Veloce have concentrated their publishing efforts on all-things automotive. Now, in a break with tradition, the company launches a new imprint for a new publishing genre!

The Hubble & Hattie imprint – so-called in memory of two, much-loved West Highland Terriers – will be the home of a range of books that cover all things animal, all produced to the same high quality of content and presentation as our motoring books, and offering the same great value for money.

More titles from Hubble & Hattie

Animal Grief – How animals mourn for each other (Alderton)
Clever Dog! (O'Meara)
Complete Dog Massage Manual, The – Gentle Dog Care (Robertson)
Dieting with my dog (Frezon)
Dinner with Rover (Paton-Ayre)
Dog Cookies (Schops)
Dog Games – stimulating play to entertain your dog and you (Blenski)
Dogs on wheels (Mort)
Dog Relax – relaxed dogs, relaxed owners (Pilguj)
Exercising your puppy: a gentle & natural approach – Gentle Dog Care (Robertson/Pope)
Fun and games for cats (Seidl)
Know Your Dog – The guide to a beautiful relationship (Birmelin)
Living with an Older Dog – Gentle Dog Care (Alderton & Hall)
My dog has cruciate ligament injury – but lives life to the full! (Häusler)
My dog has hip dysplasia – but lives life to the full! (Häusler)
My dog is blind – but lives life to the full! (Horsky)
My dog is deaf – but lives life to the full! (Willms)
Smellorama – nose games for dogs (Theby)
Swim to Recovery: canine hydrotherapy healing (Wong)
Waggy Tails & Wheelchairs (Epp)
Walkin' the dog – motorway walks for drivers and dogs (Rees)
Winston ... the dog who changed my life (Klute)
You and Your Border Terrier – The Essential Guide (Alderton)
You and Your Cockapoo – The Essential Guide (Alderton)

www.hubbleandhattie.com

First published in February 2011 by Veloce Publishing Limited, Veloce House, Parkway Farm Business Park, Middle Farm Way, Poundbury, Dorchester, Dorset, DT1 3AR, England.
Fax 01305 250479/e-mail info@veloce.co.uk/web www.veloce.co.uk or www.velocebooks.com.
ISBN: 978-1-845843-35-9 UPC: 6-36847-04335-3

Readers with ideas for books about animals, or animal-related topics, are invited to write to the editorial director of Veloce Publishing at the above address.
British Library Cataloguing in Publication Data – A catalogue record for this book is available from the British Library.
Typesetting, design and page make-up all by Veloce Publishing Ltd on Apple Mac. Printed in India by Replika Press.

Contents

INTRODUCTION

The huge, but graceful and endearing Irish Wolfhound does not, unfortunately, enjoy an especially long life. He is mature at about three years of age and may show signs of slowing down by about the age of five or six.

It can sometimes be quite difficult to define exactly what we mean by the term 'older dog.' After all, a two-year-old dog is older than a one-year-old dog, but, in this case, both would be thought of as positively youthful in canine terms. We must also consider the fact that some kinds of dogs are much more long-lived than others, and therefore, the number of years at which elderly or senior status is recognised varies from breed to breed. In some large dog breeds, such as Great Danes, Irish Wolfhounds and St Bernards, senior status is reached at about five or six years old. By contrast, a six-year-old Chihuahua – a breed that may live for sixteen years or longer – is more or less still in his adolescence at this age. Overall, the longer a dog lives, the later he achieves senior status. Nevertheless, the average age by which we assume that a typical dog has reached senior status is around seven or eight years. Determining the age expectancy, and thus the likely time at which senior status is reached, is of course much harder where crossbred dogs are concerned.

The St Bernard is another very big breed of dog with a shortish lifespan – about ten years or so. He can be prone to skin problems, tumours and hip dysplasia.

The little Chihuahua, like several small breeds of dog, can live for sixteen years or more. Chihuahuas are active dogs that will need regular daily exercise to keep them fit and healthy throughout their life.

Their heritage is not only mixed, but sometimes not even known for certain – although many experts put the average life expectancy of a typical crossbreed at about thirteen years.

Then there is our own human perception of canine age to consider. If we have owned a dog from the time it was a boisterous and mischievous little puppy, we often still consider him to be extremely youthful, even when he has reached quite an age. Provided he is still active, enjoys playing, looks trim and still does the endearing things that give each dog his individual character, nothing much will appear to have changed. Eventually, however, the signs of ageing will begin to manifest themselves, even in the fittest, leanest and most youthful-looking of dogs. There will be noticeable physical changes, such as a greying of the fur, especially around the head, and a hazy coating over the eyes, as well as certain behavioural changes, such as the propensity for your dog to want to sleep longer, run around less, and perhaps to drink more water.

As long as she is healthy, an older dog, such as this Viszla, will still enjoy playing a game with her owner, even when well advanced in years.

A new beginning

However, these are rarely signs that we need to worry about unduly. After all, if a dog lives until, say, thirteen or fourteen years of age or more, then when 'older dog' status kicks in at around seven years or so, he's only about halfway through his lifespan! Instead, we should come to regard this canine 'coming of age' as a cause for celebration and simply another phase in his life; we can think of our companion as a wise and trusted friend, instead of an unruly teenager, and look forward to our golden years together. If anything, the bond between dog and human becomes stronger at this time, giving us a chance to re-evaluate and appreciate the contribution that our dog brings to our lives. Your faithful friend will gradually make his own adjustments as he gets older, and we should be prepared to do the same. By understanding

Growing old together. This senior dog and his owner have clearly formed a close bond of friendship and trust.

what is going on at the canine level, we can help to enhance his lifestyle, improve his health, and make him an even more valued member of the family. If, for example, our dog wants to play less often, we must respect his wish, rather than try to encourage him to play, simply because that's what he always used to do. Instead, make play time with your dog a shorter, quality experience for all concerned.

There is also an important role we can play in a dog's early years to help offset the effects of ageing. In this respect, dogs are somewhat like humans: a dog that has been given plenty of exercise – both physical as well as mental – and lots of interaction with humans and other animals, will often be slower to show the effects of the ageing process. His diet throughout his earlier years can also play an important part in determining how and when the signs of ageing begin to show. An overweight dog will be less likely to actively exercise, and this in turn will mean that he may

Unfortunately, some breeds of dog – such as the the Boxer (left) and the Bulldog (above) – can be more prone to diseases in later life than others. For all dogs, however, a healthy lifestyle and diet can often help to offset problems in later years.

begin to show his age much earlier, and may even succumb more quickly to some of the diseases to which older dogs are often prone.

Choosing an older dog

So far, we have looked at the consequences of a dog changing from a puppy to a senior. Sometimes, however, an older dog comes straight into our lives without us having known him when he was younger. This may happen for a variety of reasons.

It's not unknown for someone to acquire the elderly dog of a relative who has died, or has become unable to care for the animal, for example. More likely, though, an elderly dog comes into our lives through one of the many canine adoption agencies specialising in homing stray, mistreated or abandoned animals. Many of these dogs are quite young, of course, but a number are older animals – some of which may well have found themselves in need of a new home, because of the death or incapacity of an owner as mentioned earlier. Such dogs can represent a rewarding and fulfilling experience for a new owner willing to take on an animal not in the first flush of youth. Many are well trained, and most are affectionate and gentle creatures that need nothing more than kindly reassurance and understanding in order to enjoy the rest of their life to the full, and to become your trusting and loving friend.

A modern animal shelter. Not all rescue centres may look as inviting as this one, but the staff are just as dedicated – and the residents still longing for a good home.

Rescue dogs

Rescue centres have all kinds of dogs looking for a new home. The staff at the shelter should be able to help you find an older dog that will settle well with you. Someone from the centre will usually want to visit you at home first to get an idea of your lifestyle, and also to make sure the environment is suitable. Dogs that will settle most readily are likely to be those who have lived previously with just one owner. As already mentioned, there may be a variety of reasons why owners can no longer look after their pet, which do not reflect in any way on the dog itself. On the other hand, there will be some dogs that are noisy, destructive or even aggressive, and their owners may have felt unable to cope.

Try to ascertain as much as you can about an older dog's background if possible. This is particularly significant if you are looking for a family pet, because an older dog that is unused to children could become nervous in their company and may react aggressively if frightened. It is always a good idea to arrange to take a potential dog for a walk if possible, rather than simply looking at him in the kennels. This will give you a valuable insight into his behaviour, enabling you to discover whether he responds to basic commands, and if he is accustomed to walking on a lead. You will also get a much better idea of his personality, and how he reacts to you. This can be especially critical in the case of a dog that has been badly treated. He may not take to you readily, because something about you – perhaps even your coat or voice – reminds him of his previous owner. Clearly, if you do not bond, there could be problems in the future.

Introducing an older dog to the home is rather similar to having a puppy, although you will not necessarily be faced with the problem of house-training. If this is required, however, take the dog out early in the morning, during the day, and again in the evening, when he is most likely to want to relieve himself, and then praise him when he performs, just as you would a puppy.

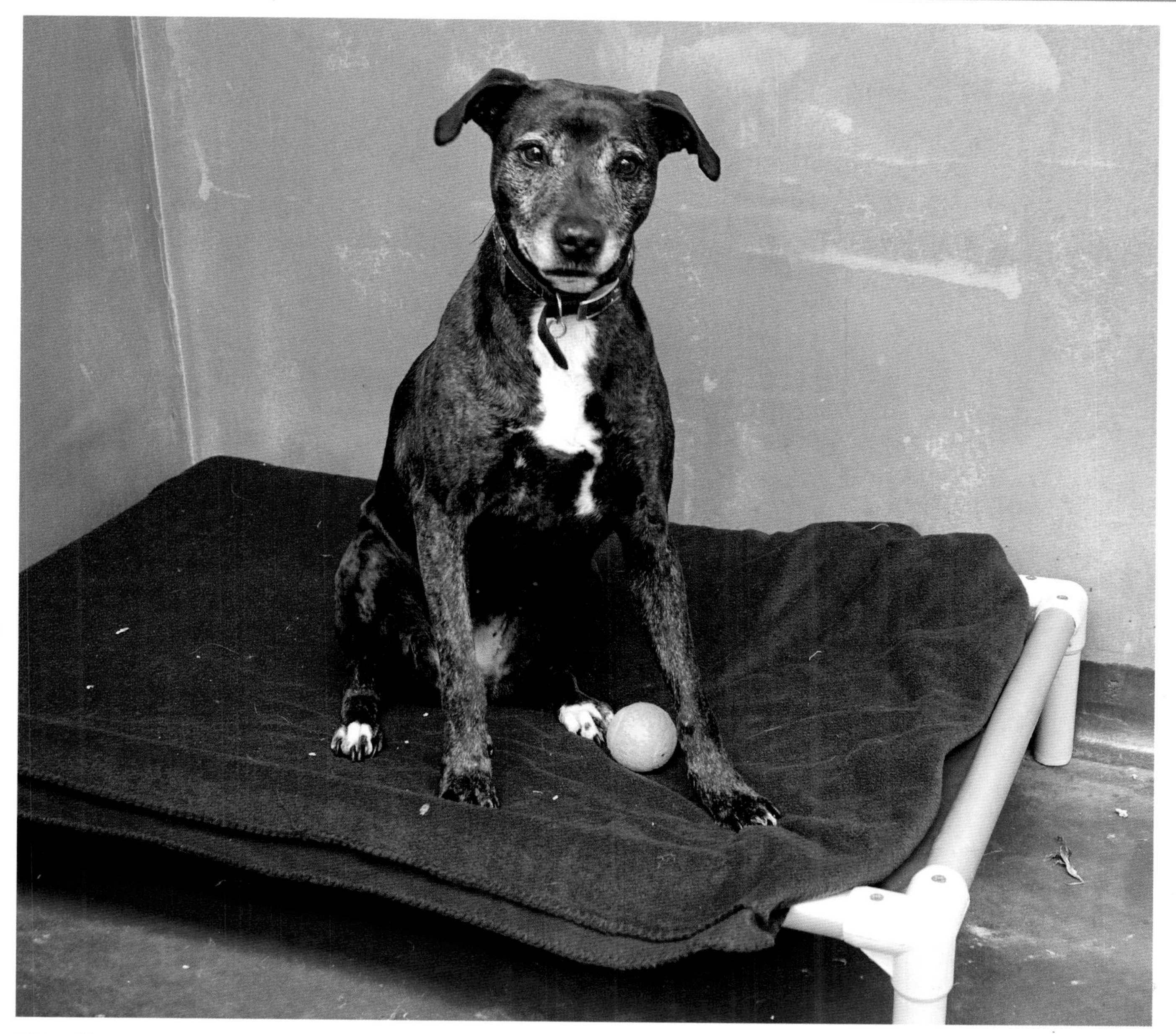

This older crossbreed looks appealing, but his true character won't shine through until he becomes settled in his new home and has bonded with his new owner.

You may discover that your new dog already knows how to make himself comfortable. Such habits may be hard to break, but if you don't want your dog on the furniture, you must firmly, but kindly, make him get down, and then praise him when he is on the floor.

Some older male dogs may seek to scent mark with urine in the home. Clean this up thoroughly, but avoid using household disinfectants, since some of these will attract the dog to soil the same site again. It's much better to use special disinfectant and de-scenting products available from pet stores. However, the best long-term solution to this problem is to arrange neutering without delay, if it has not already been done.

Older dogs are often just as likely as puppies to scratch around the home, especially at the outset; therefore, try to avoid leaving your dog on his own with the door shut, otherwise you may find the carpet damaged, especially near the door where he has been trying to get out in your absence. Even linoleum tiles may not be sufficiently resistant to prevent damage by a dog's claws, and neither is woodwork immune.

A difficult life

Once a dog has been badly treated, it can be very difficult to win his confidence, especially since you will not necessarily be aware of what frightens him. You will need a lot of patience, and must realise that he may never be as trusting as a dog that has had a normal upbringing. Sadly, some dogs pass through several homes in succession, and these are often the most difficult to re-socialise, since they can be very withdrawn and nervous by this stage. Certain breeds used for guarding, such as the Doberman and Rottweiler – as well as crosses bred from them – often have more aggressive instincts than other types of dog. Both breeds were created as guardians, so are not so much aggressive as protective and territorial. These dogs are more inclined to anticipate a threat and react accordingly, although they are very trainable; they are used by police forces, and highly valued for their intelligence.

Taking on such a dog without knowing his background can become a real test, and should never be viewed as a cheap way of obtaining a purebred dog.

Some dogs in rescue centres are likely to have been abandoned on the streets and then caught by dog wardens. Obviously, it's likely that little is known about the background of such dogs, either, or why they were abandoned. As a result of their past lifestyle, they may be seriously underweight, and could even be heavily infested with parasites such as fleas. Signs of past injuries may also be apparent on their bodies. These dogs are often rather shy and withdrawn with people, and may also show a greater tendency to wander than those that have previously lived in a home.

Just as with a young puppy, you should arrange a health check for an older dog with your vet at an early opportunity. The likelihood is that such dogs have not been vaccinated, and are therefore at risk from illnesses such as distemper.

Greyhounds, retired from the racetrack, can make excellent older pets. Keep a muzzle on them in public areas, however, to avoid them possibly chasing and injuring small dogs.

Retired racers

When their careers on the track finish, racing Greyhounds are often given new homes. Their gentle natures mean these hounds can make ideal companions, although they will need some time and understanding to make the necessary adjustments from kennel to domestic life. Because they have spent the first few years of their lives in kennels, Greyhounds may need house-training.

A significant point to bear in mind when exercising a retired Greyhound in any public area is that it will have spent its former life chasing small furry lures around a track, and may respond in a similar way when meeting up with small dogs such as Yorkshire Terriers. The smaller dogs will not appreciate the danger, and are liable to be caught and even killed by the Greyhound. This will be exceedingly distressing for all involved, and can also leave you facing a large bill for compensation from the owner of the other dog. The simple

Lurchers are a crossbreed – often between a Greyhound and a Collie, for example. An older Lurcher like this one can also make a delightful and good-natured pet.

solution is to exercise your Greyhound off the lead only when he is wearing a muzzle, which can be purchased from most pet stores and should not cause your dog any great discomfort, although it is important to ensure that it fits correctly. In spite of their reputation as canine athletes, Greyhounds only need short bursts of exercise, since they are sprinters rather than long distance runners.

Before letting any dog off the lead, you must be certain that he will return to you when called, which is why you should not allow your new companion to run free for several weeks after you have acquired him. Obviously, the time frame will depend on the individual dog and his background; if he is already well trained, then he simply needs to become accustomed to you, whereas a long-term stray will require much more training. Your local dog-training school should be able to assist – details of the secretary can usually be obtained from your vet or local library.

It's best not to encourage your dog to sleep in your bedroom, however comfortable he looks!

Behavioural problems in the rescue dog

One of the most disturbing problems in the rescue dog is likely to be coprophagia, a condition where the dog eats his own faeces, a problem that occurs most often in dogs that have been kennelled or who have lived in fairly unhygienic surroundings. It is possible to correct this behaviour by preventing the dog from having access to his faeces. In severe cases, veterinary help may be required: there are drugs available that will cause the dog to vomit after consuming his faeces, or give them a highly unpleasant taint.

Another problem is destructiveness, where the dog damages the home in your absence, and may also be linked with persistent barking when he is left alone. Such types of behaviour can be a reflection of separation anxiety on your dog's part. There may also be something in his past underlying these activities, particularly if he has been subject to neglect or left confined for long periods on his own. The issue of separation anxiety is discussed in the chapter entitled *Physical and behavioural changes*, along with strategies for dealing with the problem.

Some dogs may also bark when left at night. While this could obviously be indicative of intruders, you do need to be careful that your dog does not use this ploy to get you out of bed for reassurance on a regular basis. This difficulty is most likely to arise if you allow your dog to sleep in your bedroom for the first week or so, and then expect him to remain on his own elsewhere in the home afterwards. It is much better to start as you mean to continue, and thereby encourage the dog to sleep in his permanent night-time location.

Be sensitive

Other potentially problematic behavioural traits may emerge as your new dog settles into his surroundings. Unless you are familiar with his background and history, be careful, for example, when taking objects from him, since this could result in you being unexpectedly bitten. This applies particularly if the dog has not been trained to drop on command. Rather than place your hand around the object in the dog's mouth, put your left hand across the muzzle and use the other to prise down the lower jaw, causing the dog to release the item. Even if he tries to snap, there is less likelihood of any injury, since you are restraining his jaws.

Dogs who react by snapping if their toys are taken from them are by no means vicious and, in other situations, their behaviour may be impeccable. The reason for such a reaction almost certainly resides in their past, coupled with poor training. In time, and with suitable encouragement, you may well be able to overcome this vice – although this does not mean that your dog will behave in an equally co-operative way with other members of the family, unless they have been closely involved in his care. If you encounter real difficulty relating either to the dog's training or his general behaviour, it may be worth contacting an animal behaviourist for help. Furthermore, make sure children are aware that they should not attempt to take away items from the dog on their own, but should instead seek adult help.

Pseudopregnancies

Be very careful with a dog suffering from a pseudopregnancy. This hormonal condition arises just after the time that a bitch would normally give birth – some 63 days after her last season – and may even cause her mammary glands to swell and produce milk, even though she did not mate. In the absence of puppies, she transfers her maternal affections to her toys, and will become very possessive towards them. Pseudopregnancies often recur repeatedly after each season, so the simplest solution is to arrange for her to be spayed in order to prevent the problem arising again.

PHYSICAL AND BEHAVIOURAL CHANGES

Just as with humans, a dog's appearance changes as she gets older. Compare the somewhat sedate and stocky senior Springer Spaniel on the left here with the much more youthful example of the breed.

Changes in a dog's appearance and behaviour do not just happen in her senior years, of course. Ageing begins at birth, for all of us, including dogs! However, the changes we discern when we see our dog grow from a puppy into an adult tend to be those associated with strength, vigour and vitality, whereas those that come on during the advancing years tend to reflect a more stately body and mind. Things will begin to be done at a slower pace; a little more time may be needed before your dog is ready to 'get going' in the morning; and her periods of sleep will become longer and deeper, and probably more frequent as well. And although it may not always be readily apparent, this loss of physical condition is often accompanied by a deterioration in mental faculties, which can sometimes give rise to emotional changes in a dog, too.

It sometimes takes an older dog longer to become fully awake and ready for the day ahead. Let your senior get going at her own pace ...

Common signs of ageing

- *Changes in the coat. With age, the coat begins to grey, especially around the muzzle, eyebrows, and face in general. It may also look a little thinner and less sleek*
- *You may begin to see and/or feel small, softish lumps on your senior dog's body. Although these can indicate a cancerous tumour, in many instances they are simply benign fatty lumps known as lipomas, or wart-type growths called papillomas. Get your vet to check out all lumps, however*
- *Stiffness in the limbs, especially after sleeping and after exercise, is more common in older dogs*
- *Older dogs generally start to take life at an easier pace, so you may find your dog has less enthusiasm for playing and walking*
- *Accidents around the house are another sign that your dog's bladder is not as efficient as it once was*
- *The eyes and ears may not be as sharp as they were, which means your dog may sometimes seem to ignore you when called (simply because she hasn't heard you), or may show signs of failing eyesight*
- *Older dogs may begin to show an intolerance of loud noises (such as fireworks or thunderstorms, when they didn't previously), and may exhibit other signs of anxiety (for example, when left alone in the house)*
- *Your dog may want to drink more frequently*

It is not unusual for older dogs to drink more frequently. There are several reasons why this may happen, including kidney problems and diabetes. Drinking too little water can also be a problem, since this can cause dehydration. Consult your vet if either of these happen.

This senior is clearly overweight, as can be seen by comparing her body shape with the middle diagram of the dog body shape profiles below.

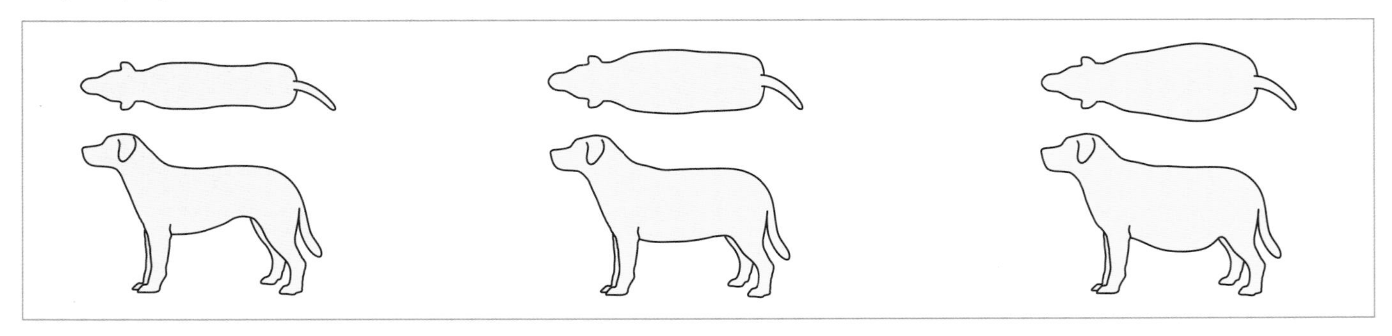

Side and from above profiles of a dog with normal body weight (left), a dog that is overweight (middle), and one that is obese (right). Unless there are specific reasons for it, a tendency towards obesity or being overweight should be addressed, since these conditions are detrimental to good canine health.

When a dog ages, her coat becomes less sleek. However, regular, careful grooming, and perhaps the occasional supplement, can help keep it in good condition.

The coat

One of the most obvious physical signs you may notice as your dog ages is a change in her coat condition. The coat is, in many ways, a mirror of the condition of the underlying skin; the individual hair follicles that make up the coat have their origin in the deep layers of the skin. Special muscles in the skin, called erector muscles, pull on the hair follicles to raise the hairs, and the sebaceous glands in the part of the skin known as the dermis keep the hairs lubricated, by secreting an oil called sebum. With age, the skin loses its pliability, and the capacity of the oil-producing sebaceous glands also reduces. These changes can cause the coat to look drier and out of condition, and can also cause the fur to lie less sleekly. The coat may begin to look thinner, as hair regeneration slows.

Another obvious sign of ageing is greying of the fur, especially noticeable in dogs with dark fur around the muzzle and other parts of the face. Like humans, this is brought about when the cells called melanocytes (that give individual hairs their colour) produce less pigment, and the hair then appears colourless, or 'grey.' There is little you can do about greying in your senior – it's just part of the ageing process – but there are various coat conditioners on the market that can help to replace some of the oils that are lost naturally with age, making the

With age, greying of the fur occurs around the face and muzzle. This is especially easy to see in dark-coated dogs like this black Labrador.

With advancing years, the eyes of a dog become cloudy and may take on a bluish tinge, as seen here. A whitish coloration of the eyes is more likely to be associated with the development of cataracts.

coat appear sleeker and in better condition. Some of these are available in the form of a cream that you work into the fur, and others are sprayed onto the coat.

The senses

You may also notice changes in your dog's eyes. The lenses of the eyes may begin to appear cloudy, which, often, is related to a condition that doesn't actually affect the ability to see, although it can sometimes be associated with the onset of age-related conditions such as cataracts. Changes to the elasticity of the lens mean that your senior can experience nearsightedness, which may be evident when she is trying to focus on and catch a ball, for example. There can also be a reduction in the ability to see in the dark, or even in bright light. Most of these are things that you can make allowances for, so that your dog continues to live life to the full. For example, when playing with a ball, try rolling it along the ground instead of throwing it in the air; for a dog that isn't seeing quite as well as she used to, it can often be easier to keep track of a ball in this way. When out walking, be especially aware of any hazards, such as wire fences, that she may have trouble seeing when running at speed. But don't despair, even if your dog's eyesight is really poor; the power of her other senses, such as smell, means that dogs can compensate remarkably well when partially or even totally blind, providing they have learnt to adjust to the situation and are in familiar surroundings.

Your dog's hearing may also deteriorate with age. Often, this is caused by nerve damage, and may be a hereditary condition in some breeds. Deafness can also be linked to a hardening of the bones in the middle ear – a condition known as otosclerosis, which, again, is age-related. Sometimes, apparent deafness may be due to an ear infection, a build-up of earwax, or to some foreign body inside the ear.

To help prevent separation anxiety, get your dog happily settled before you leave the house. A favourite toy, such as an old slipper, may help to reassure her.

Separation anxiety

An ageing dog with diminishing vision or hearing may exhibit anxious behaviour when she is left, a condition often known as separation anxiety. Changes in the dog's normal set routine can make matters worse. The anxiety may begin to show even before her owner leaves, and may manifest itself in bouts of howling, urinating and defecating, or even by destructive behaviour while she is on her own.

Dealing with separation anxiety

- *Try altering the manner of your departure. Dogs pick up on visual cues, such as any routine you may go through before you leave. Can you actually avoid some of these altogether – like not jangling your car keys, for example? At weekends, consider going through your 'getting-ready-to-go-to-work' procedure, but instead of actually going, spend time playing with your dog instead – even if you are wearing work clothes!*
- *Make sure your dog is calm before you do go out. Initiate this by helping her to relax and sit quietly in your presence. A tense dog is likely to become even more stressed when she is left on her own*
- *Leave your dog in a good environment. You know what she likes best, so be sure to put her bed in a favourite place, and allow her some freedom to roam about, or to look out of the windows if she likes doing this. Try leaving a treat that will occupy her attention while you are gone. Leaving the radio on can help to give dogs some other human 'company' in your absence*
- *Make your absences very short initially. You can even try simply leaving your dog in one room while you close the door and go into another for a short time. (This could be for a few minutes or even just a few seconds, depending on your dog's reaction.) Each time she remains calm whilst you are away, praise her when you return, and don't scold her when she doesn't behave as required; simply ignore any inappropriate behaviour. Gradually increasing the time your dog is left, and then introducing a short period during which you are actually outside the house, should eventually bring about a change in behaviour, even though this may take some weeks to achieve*
- *Have someone visit while you are out. It sometimes helps to have a friend, neighbour or relative come to the house and spend some time with your dog. This reduces the time that she is left alone, may reduce her anxiety, and can give her the opportunity to relieve herself outside and take a little exercise*
- *Keep your departure and return low-key, so that your senior doesn't associate this behaviour on your part with being left alone*
- *Discuss the problem with other dog owners who may have encountered the same issues and have other ideas for you to try. An Internet discussion forum is a good place to look. Get some advice from your vet or an animal behaviourist if necessary, too*

Some companies allow staff to take their dogs to work – an ideal remedy for separation anxiety!

Like several other kinds of very large breed, the Great Dane is not a long-lived dog, with a life span, generally, of between seven and ten years. He may show signs of ageing at a time when many smaller breeds are still quite youthful-looking.

Aches and pains

As dogs age, so their internal organs also begin to show signs of wear and tear. Bones become more brittle, and cartilage (the tissue that caps the ends of bones at the joints) erodes. When the cartilage wears away, the ends of the bones can rub against each other painfully, producing the symptoms of one of the various forms of arthritis. The hereditary disease known as hip dysplasia can also seriously affect the mobility of some breeds by the time they reach their senior years.

Muscles begin to lose their mass and tone, resulting in more difficult, or even painful movement, and this reduces agility and the desire

In her senior years, your dog may try to avoid lying on hard surfaces. This Labrador has chosen the carpet rather than the wooden floor, and who can blame her!

to run around and play. This general stiffness can manifest itself in many ways: you may notice that your senior takes longer to get up and lie down than previously; what was once done at a trot is now done at a steady plod; and staircases and settees need to be negotiated using special strategies, or even a helping hand from you. Maintaining your dog's correct body weight, adhering to an appropriate diet, and the right type of exercise can all go a long way towards offsetting the effects of musculoskeletal problems, however. There are also ramps that can make clambering up into a vehicle an easer operation when travelling (see the chapter entitled *General care*).

Many older dogs experience more difficulty in tolerating extremes of temperature. You might begin to see a reluctance to go outside, if it is wet or cold, coupled with a greater keenness to seek the shade on hot days. Make sure you notice this behaviour, and respect her wish to stay within her comfort zone. Provide a warm blanket for your dog to lie on, rather than a tiled or wooden kitchen floor, and erect some form of shade in the garden in sunny weather, if no natural cover is available.

Give your older dog plenty of time to perform her toilet duties. Activities like this can take longer than they did when she was a youngster.

Body functions

As well as the various musculoskeletal changes mentioned above, other changes also take place inside a dog's body as she ages. The internal organs – such as the heart, liver, kidneys and digestive system – may lose some of their efficiency. The immune system, whose job is to protect against invasion by foreign bodies, may also decrease in function, becoming less efficient at warding off attacks from bacteria and viruses. The digestive system of an older dog may not be able to process food as efficiently as previously, and it pays to consider both the type and amount of food that you feed to a senior. Adjusting the amount of calories that your dog eats when she gets older will help to avoid obesity, a condition that can seriously affect health and life expectancy.

An older dog that begins soiling around the house may have an underlying medical problem, such as colitis, a bladder infection, or a disease of the kidneys. There may come a time, due to some physical disability, when your senior starts to find it difficult to get outside to relieve herself, so it may be necessary to improve access to the garden in these circumstances. Some kinds of food can make it difficult for a dog to defecate, and your vet may advise you about a change of diet that can overcome this problem.

As a rule-of-thumb guide, a healthy dog will normally drink about 1fl oz (30ml) of water per pound (2.4kg) of body weight every 24 hours, but older dogs are often prone to drinking more water than previously. This problem may be related to conditions such as Cushing's disease or diabetes, for example, but it is also likely to be a sign that the kidneys are beginning to function less well. Your vet may recommend a change of diet to help combat the effects of kidney disease. Such a diet is lower in protein, as well as in sodium and phosphorus. Kidney disease can also be a cause of bad breath in a dog, although other reasons include mouth infections, oral cancers, gum disease, and a bacteria-laden tartar build-up on the teeth.

Pyometra, an infection of the uterus, can affect young and middle-aged female dogs, although it is most common in older animals. The signs of pyometra vary according to whether or not the cervix is open. (Spayed dogs won't have this condition, because it's a uterine infection. In fact, spaying to remove the uterus and other reproductive organs is a method of dealing with the condition when it presents.) The cervix is the neck of the uterus, which usually remains tightly closed, and only opens when the female dog is about to give birth to pups. When it is closed, the pus caused by pyometra won't discharge or be visible, but, if it is open, there will be signs of pus draining from the uterus; this may also be visible on bedding or furniture where your dog has lain, or on her fur and under the tail. If the cervix is closed, any pus that forms is unable to drain away to the outside; instead, collecting in the uterus, where it can cause a distension of the abdomen. The bacteria in the discharge release toxins, and these enter the bloodstream where they can cause problems such as vomiting, diarrhoea and

Excessive panting may be a symptom of stress in the older dog.

listlessness. The side effects of pyometra can also affect the kidneys, again, causing the dog to drink and urinate excessively. If any of the described symptoms are noticed, see your vet as quickly as possible.

Changes in behaviour

In addition to the separation anxiety mentioned earlier, an older dog may increasingly exhibit other behavioural changes. For example, a once-placid and well-mannered dog may show signs of aggressive behaviour in seniorhood. However, this aggression may simply be due to a physical problem. Sore bones and muscles, or a dental problem, may be causing your dog pain, and making her short-tempered and even frightened. It may also be the case that your dog is getting hard of hearing or not seeing too well, and becomes frightened when you or someone else in the household inadvertently startles her. An older dog can also show signs of stress if her normal routine is suddenly changed – for example, if you move house, or even just change the position of the furniture in a room. Leaving the dog alone in the house for longer periods than usual, or introducing another dog into the family can also be unsettling experiences for an older dog. Many of these problems can be overcome by sensitivite management of your dog's environment, the application of common sense, and also by observing what factors seem to be making your senior react in this way.

Stress can manifest itself in many ways – excessive panting, an inability to settle, and the propensity to follow you around everywhere can all indicate that your dog is not happy with the situation. Mood changes can also be the result of an endocrine imbalance, or due to the condition known as cognitive dysfunction syndrome (see later).

Another sign that your dog may be stressed is if she resorts to barking, whining or howling. This can occur as a result of separation anxiety (see page 27), as well as being used as a strategy to try and get your attention. A form of therapy called remote correction attempts to modify this behaviour by distracting the dog when she vocalises. The best way of doing this is to toss a tin filled with pebbles, or rice, in her direction, which will slightly startle her because of the unfamiliar noise it makes. Because your dog won't like the sound of this, she will begin to associate barking with a disagreeable surprise (the noise), so will stop barking. How quickly this happens really depends on your dog; some respond quickly, and others take longer. If you try this, don't let your dog associate you with the correcting behaviour, however; toss the tin when her attention is directed away from you, while she is occupied by barking. Alternatively, you can have an accomplice, out of your dog's immediate view, who slyly tosses the tin while you have her attention.

Stress can also manifest itself in other forms of behaviour: your senior may develop a change in her eating habits or appetite; she may also begin to find it more difficult to settle at night, pacing about the house, and may show signs of struggling to urinate or defaecate. Your vet should investigate any of these signs.

Noise phobia

Fireworks, the rumble of thunderstorms, high traffic volume, and other loud sounds can make your senior dog extremely anxious and fretful. There are several reasons why this might be so: she may have less mobility, making it harder to remove herself from what she perceives to be the source of the noise; she may be less able to manage stress than when she was younger; and she may be suffering from cognitive dysfunction syndrome (see later). There are various therapies available to treat noise phobias, including medication that can calm her. Another method is counterconditioning. With this treatment, the same sound that causes the phobia is played to the dog initially at a low level, and the sound then gradually increased over several days or weeks. A reward is given each time the dog displays no fear of the noise. Even without using such measures, there may be practical things you can do to alleviate the problem. Closing all windows and drawing curtains will help to reduce the level of noise. Talk gently to your dog to calm and reassure her, or try offering a distraction, such as playing her favourite game. If possible, do not to show any anxiety yourself – even if you are afraid of thunderstorms as well! If you do so, you will simply transmit your fear to your dog and reinforce the idea that she should be afraid.

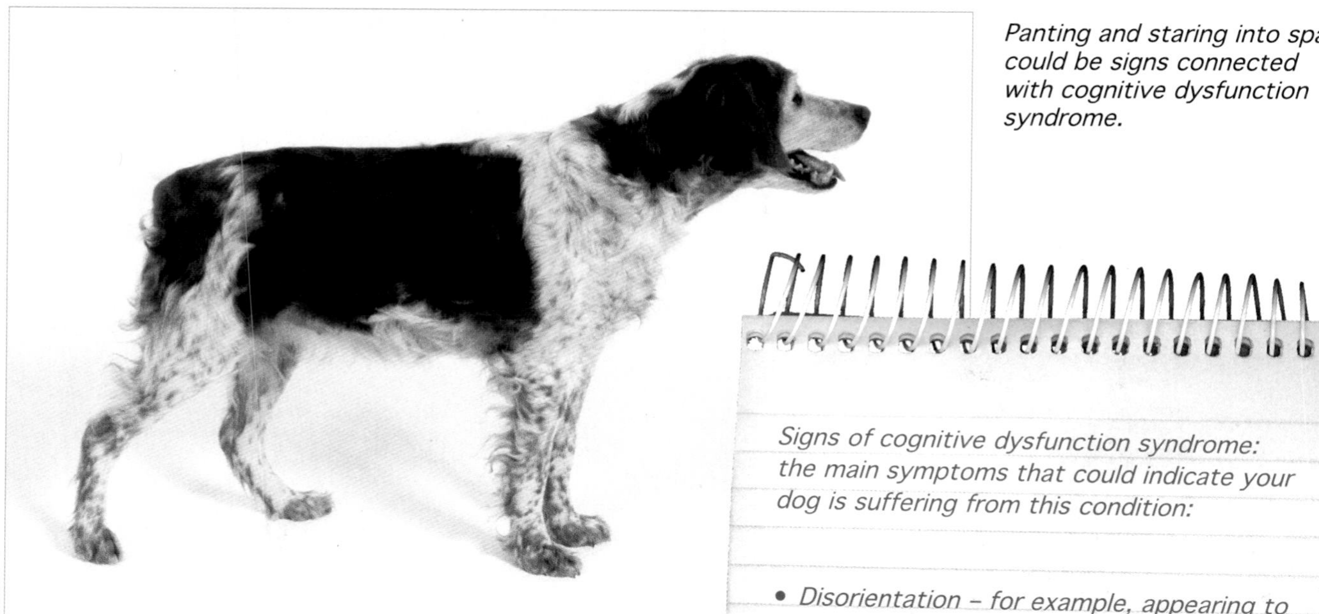

Panting and staring into space could be signs connected with cognitive dysfunction syndrome.

A condition that can afflict some older dogs is cognitive dysfunction syndrome. When your dog ages, the brain will age, too, which can result in a form of dementia or senility similar to the condition Alzheimer's disease that is seen in humans. It is calculated that more than half of all dogs over the age of ten years will experience at least some of the symptoms of cognitive dysfunction syndrome.

Because some of these signs may also indicate other diseases, your vet will need to carry out a series of diagnostic tests to try and ascertain whether or not your dog is suffering from cognitive dysfunction syndrome. Unfortunately, as in humans with Alzheimer's disease, there is no cure for this condition. Your vet may recommend treatment with a drug such as Anipryl, though, to help offset some of the symptoms. Even though cognitive dysfunction syndrome is not preventable, there are plenty of games and other stimulating pastimes that you can undertake to try to keep your dog mentally active as she enters her senior years (see page 51 of chapter entitled *General care*).

Signs of cognitive dysfunction syndrome: the main symptoms that could indicate your dog is suffering from this condition:

- *Disorientation – for example, appearing to become 'lost' in familiar surroundings such as the garden, or being 'trapped' behind furniture*
- *Failure to recognise familiar people, or to respond to familiar commands or cues*
- *Showing anxiety for no apparent reason*
- *Showing no interest in familiar activities such as playing or being stroked*
- *Showing confusion or staring into space*
- *Unusual behaviour, such as barking or whining without reason, circling or shaking, or wandering about at night*
- *Appearing aggressive for no apparent reason*
- *Accidents in the house that cannot be attributed to any known health problems*

GENERAL CARE

Just because your dog is now more elderly, it doesn't mean that you need to fuss over him endlessly and treat him like a geriatric. However, in the same way that an elderly person might begin to make adjustments to their lifestyle, it also makes good sense to consider introducing changes in the daily routine of your senior dog.

The changes don't need to be dramatic or sweeping – after all, a dog isn't young one day and then suddenly old the next – but should be introduced gently, and increased gradually as time goes on. Some dogs, having been accustomed to an active and healthy life, hardly seem to want to slow down at all, despite the advancing years. Your vigilance, and your understanding of your senior dog's needs and behaviour, will play a big part in determining how much quality time is available to him in his latter years.

Active breeds, such as this alert-looking Border Collie, may show little sign of wanting to slow down, even as they enter their senior years.

A warm, waterproof coat for keeping out the weather.

The daily routine

A dog is especially likely to appreciate a regular daily routine as he gets older, and life can also be made easier if you give some extra thought about your dog's needs at the same time. For example, instead of wanting to get up and about as soon as possible in the morning, your senior dog may now want a lie-in (assuming his bladder will allow it!). Therefore, be patient, and don't rush your dog out of his bed and into the garden until he shows signs of wanting to go outside. There may be some stiffness in his joints on first awakening, so allow time for some loosening up to take place before you take him out for a walk. Some older dogs are simply more reluctant to go out in inclement weather and need a little encouragement. Wearing a warm, waterproof dog coat can be beneficial by helping to keep out the cold and damp. (See also the chapter entitled *Special care*.)

An older dog may become less tolerant of boisterous children wanting to play with him and pull him about, and may even resort to a lip curl or a growl if provoked too far. This is often because his joints may now be slightly painful, in which case he

As well as a bowl of water in the house, an older dog may benefit from having an extra bowl of clean water outside.

should always be handled carefully and gently for this reason. Avoid this situation by explaining that your dog is now older and needs to take things at a steadier pace, and must be allowed more time to himself if he wants.

Your dog may also be more easily startled by loud noises, and might get a little confused from time to time. These are all signs of an ageing body and mind, and relate to the fact that the senses have become a little dulled, and that aches and pains are more commonplace. Therefore, your dog will need more patience and understanding and a slightly quieter environment. If you decide to rearrange all the furniture in a room, do it bit by bit, so that this major upheaval in his living space doesn't confuse your dog – this is particularly pertinent if his eyesight is beginning to fail. Make sure your dog's bed is placed in a warm, draught-free part of the house, out of the main thoroughfare, so that it is a comfortable and safe retreat whenever a bit of peace and quiet is needed. An older dog may become more anxious when separated from his owners, so if possible try and avoid leaving him for long periods.

If your dog becomes unable to go through the night without relieving himself, he will need retraining so he can use litter trays. Make sure these are placed on a surface that is easy to clean, however, as here.

Older dogs often begin to suffer from kidney problems, so make sure that plenty of fresh water is always available. An older dog with stiff joints will appreciate his food and water bowls being raised up slightly on a suitable support, so he can reach them more easily. There are special stands that you can buy for this purpose. Incidentally, the extra intake of water may mean that more frequent comfort breaks are required, so keep an eye open for the warning signs of wanting to go outside to help prevent accidents in the house.

Diet is an especially important factor where the senior dog is concerned. Stiffness, aches, and pains may reduce a dog's interest in exercise, but, quite often, the dog continues to receive the same amount of food as when he was more active – which can quickly result in obesity. This condition can lead to other harmful afflictions such as arthritis, diabetes and heart disease: an overweight dog puts more strain on his muscles, joints, and heart. If your senior dog is showing signs of obesity – for example, if his body shape is changing and his ribs are becoming harder to feel – the first course of action is to change his food intake. This usually means reducing the amount, but also feeding him the right kind of food. It does not mean cutting out the essential nutrients that make up a balanced diet, such as proteins, carbohydrates, fats, vitamins and minerals, but it does mean offering them in a form that contains fewer calories and more fibre, is more balanced, and more appropriate to the age of your dog. See the chapter entitled *Feeding the older dog* for more details about senior diets.

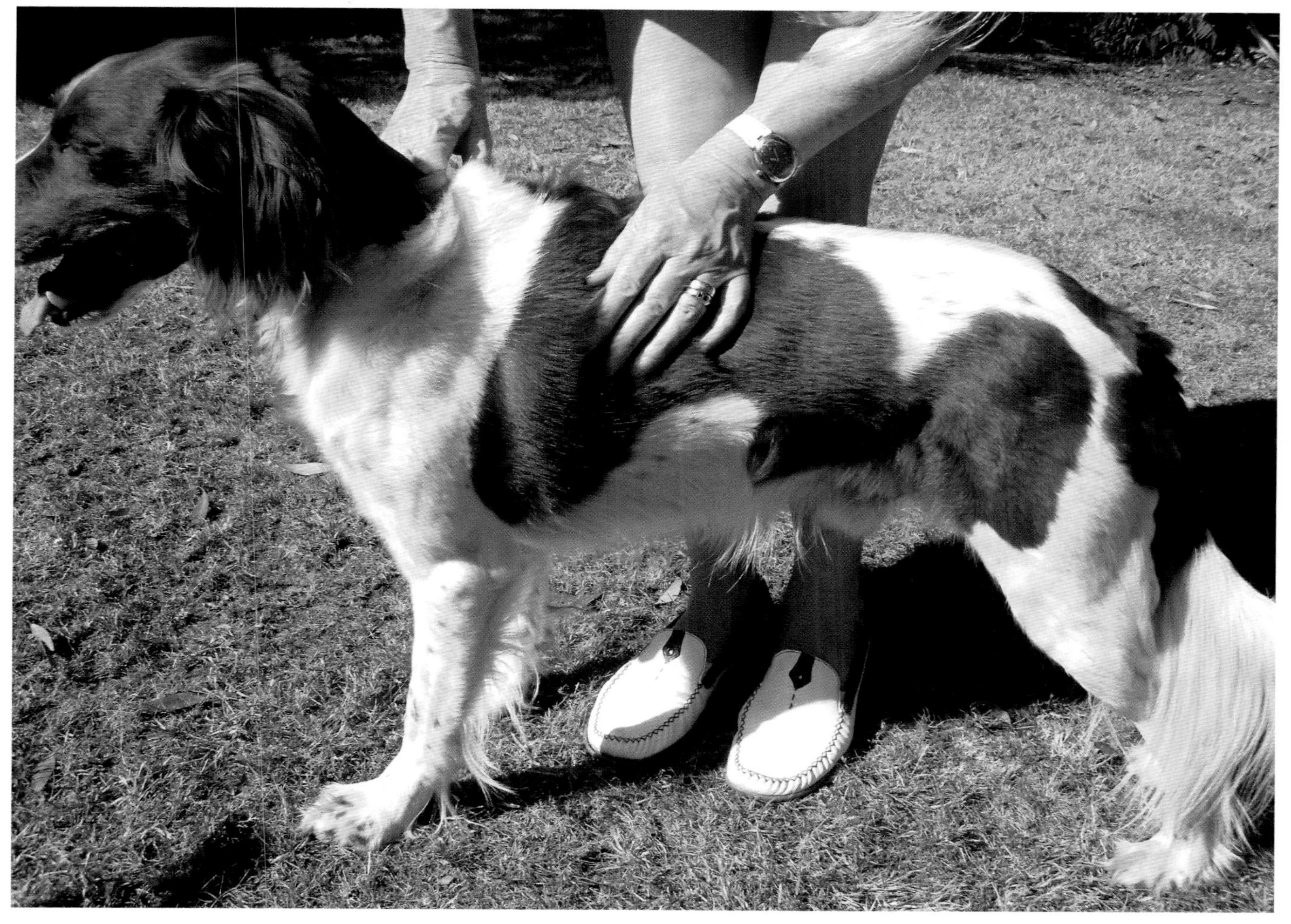

Get your dog accustomed to having his body checked regularly for any abrasions, lumps, or foreign bodies.

Grooming and checking

Grooming and checking your dog over are tasks which should be performed regularly on a dog of any age to ensure he stays in top condition. However, it is even more vital that you carry out this job where a senior dog is involved – even if a certain amount of canine crabbiness and reluctance on his part mean that you may sometimes have a bit of a battle on your hands!

An older dog's skin will become less supple, and the coat will become drier. There will also be a greater tendency for the skin to develop cysts, sores and other conditions. Once a week, therefore, carry out a thorough inspection by

At the same time, check his paws for cuts or sores.

feeling all over the body for any unusual lumps and bumps, cuts, embedded thorns or other ailments. This is also a good way to feel for parasitic ticks that dogs often pick up in the warmer months of the year; these small arachnids burrow their heads into the dog's skin and suck the blood. A good remedy for removing them is to first smear the tick's exposed body with petroleum jelly sold as Vaseline®. This will block its breathing apparatus and cause it to loosen its grip in the dog's skin, when it can then be carefully removed. The best method, however, is to use a specially-made device called an O'Tom tick hook, which completely lifts out the tick. Whichever method you use, it is vital that all of the tick is removed otherwise infection could occur if part (usually the head) is left behind. This kind of examination will also reveal anything that could be a tumour, and if you find anything suspicious, you should get your vet to take a look at soon as possible.

Brush your dog's coat daily, carefully removing any knots or tangles, and looking for the presence of little dark specks that could indicate fleas. You may even spot a flea, although this is not easy, since they are tiny and move fast. The little dark specks are, in fact, the fleas' faeces. The presence

Brushing the coat keeps it in top condition, and also stimulates the skin to help circulation.

Part the fur and check the skin, especially around the head, ears and tail base, for the presence of flea dirt or even fleas.

of flea dirt can be confirmed by placing some on a moist white paper towel; ordinary dirt will leave the paper unchanged, whereas the paper on which flea faeces is placed soon turns a reddish colour, due to the dissolved blood in the faeces. See the chapter entitled *Special care* for how to deal with fleas on your dog.

Grooming is usually an enjoyable bonding experience for you and your dog, and doing it will also help to stimulate his blood flow and keep the coat healthier by removing dead hair and dirt, and distributing the body's natural oils. However, an older dog probably won't appreciate being brushed or combed too vigorously, so be gentle, though thorough. Don't forget the ears (including inside the ears) and tail. A coat conditioner will help to keep your dog looking healthy. You will still need to bathe your dog from time to time, using only a shampoo designed for dogs. Make sure that the shampoo doesn't get into his eyes or inside the ears, and be sure to rinse the coat well with warm, clean water.

There are other tasks you should carry out when you check and groom your dog's coat. First, check the eyes by making sure that there are no problems such as sore areas around the

Check under the tail for any signs of discharge, sores or compacted faeces.

After bathing your senior, allow him to shake to remove excess water before drying with clean, soft towels.

Carefully clean around the eyes with a moistened soft tissue or clean cloth.

lids (another place where ticks like to latch on!). Carefully wipe away any 'sleep' or discharges from around the corner of the eyes with the edge of a moist cloth or a little moistened cotton wool. Lift the earflaps and look inside. Again, check that there are no cuts, sores or foreign bodies. Be aware of any discharges or unpleasant smells in the ears which may need investigation by a vet. Inflammation, soreness or crusty discharges could indicate an ear mite infection, for example.

Last but not least, check the mouth. Talk calmly to your dog and place your hand over the muzzle and lift up the sides of the mouth so you can inspect the teeth, especially where they meet the gum line. Young dogs have smooth white teeth, but these darken with age. An older dog's teeth may also show signs of plaque or tartar; this usually takes the form of a white, yellow or brown coating on the enamel. If the build-up is only slight, you may be able to brush this away with a toothbrush and toothpaste designed for dogs, but, in the case of severe deposits, a vet may need to descale the teeth. Plaque can lead to gum infections and is one of the causes of smelly breath that often seems to be a feature of an ageing dog. More importantly, oral bacteria can travel to the heart, so good dental hygiene is vital. Offer your dog one of the proprietary products such as dental sticks or

Check inside the ears to make sure they are clean, odour-free and healthy-looking.

sterilised bones that are designed to keep his teeth clean as he chews or gnaws on them. Likewise, dogs that have a dry food diet are less likely to get periodontal problems, because wet food has a tendency to stick to the teeth.

Many dogs will not be keen on having their teeth cleaned with a toothbrush and paste – especially if they are unaccustomed to the practice – so be prepared for a protest! Only clean the teeth gently, using special canine toothpaste, and stop if your dog is getting really resentful or stressed. It may be best to introduce the process by first wrapping a little gauze around your index finger, and then gently rubbing the teeth with a small amount of toothpaste smeared onto the gauze. Opt for a gradual approach, increasing how much you do a little at a time so that he becomes used to the procedure.

Gently lift the skin around the mouth and check the teeth.

Ensure that your senior dog can get around the home safely. Note here the clear, open aspect of this living room that makes it easy to walk about without the need to negotiate obstacles. Access to the garden is also unimpeded.

The senior environment

As well as keeping your dog's coat in good condition, it's important to pay attention to his bedding, which may become dirtier sooner than it did. Grubby or soiled bedding will not only make your dog a bit smelly, but the odours the bedding gives off can also permeate around the house. Perhaps your dog has a rigid plastic-type basket with soft bedding. If so, remove the bedding on a regular basis – perhaps weekly – and put it through the washing machine (have clean spare bedding available to put in the bed in the meantime). At the same time, brush out any dust and hairs that may have accumulated in the basket, and give it a wash and rinse before drying it thoroughly. Other common types of dog bed include beanbags and soft-sided bed bases with a removable 'mattress.' Beanbags usually have an outer casing that can be removed for washing, and most other soft beds are designed to be put through the washing machine. In between washes, the bed can simply be vacuumed to keep it clean. An extra blanket or two in the bed will help to make your dog warmer and more comfortable, too.

Most dogs, even older ones, are remarkably surefooted, but when your dog starts to become a little less stable on his legs there are a few ways you can help him get about more easily. Bare, highly polished floors are inevitably slippery, so it may be worth cutting back on the buffing here. Alternatively, you could lay a narrow rug or some other non-slip surface down as a 'path.' Your dog will soon work out that this is the best route to take. Many long-coated dogs have copious hair growing around the pads on their feet which, if carefully cut back a little, will help the pads to make firm contact with the floor to ensure a better grip. Make sure, too, that his claws aren't too long.

Introducing a younger dog

How your senior will react to the introduction of a younger dog into the home depends partly on the dog itself, but also on how you manage the situation. There are pros and cons with this venture: a lively young dog can give some older dogs a new lease of life, and provide some unique canine companionship; conversely, some senior dogs, set in their ways, can become extremely resentful of this intrusion into their lives, coming as it often does when they are at the least tolerant and adaptable. If you do introduce a younger dog, make sure you do so while your dog is still relatively active and in full control of all of his faculties.

Should you decide to bring a new dog into the home, make sure your older dog gets extra attention, at least to begin with, and is treated with the seniority he deserves. This can include your senior being fed first. At the very least, he should not have to compete for food with the younger dog, and it may be necessary to feed them separately, until certain rules are established. While making sure that your senior dog feels important, loved and reassured, it is also important to make the youngster feel welcome and at home. If you do this correctly, in time they should both settle into a happy relationship. Above all, don't skimp on the amount of quality time you all spend together.

When introducing a younger dog to an older one, it's often a good idea to let them get acquainted by first allowing them to meet and have a good mutual sniff on 'neutral' territory.

Your pet will probably still enjoy playing with toys well into the senior years, which is also good for exercise and mental stimulation. Ideal toys include a soft ball (such as a tennis ball), a soft, squeaky toy that can help to exercise the jaws as it is squeezed in the mouth, and some flexible toys that can be chased and shaken, and used for gentle tug-of-war games.

An older dog may be quite content to exercise in the garden.

Exercise

An older dog will still require exercise; in fact, an appropriate amount of exercise is essential for helping to maintain canine good health at any age. The advantages of fitness are numerous, and not that different from the ones enjoyed by humans who remain physically active. These include: helping to maintain muscle tone and preventing muscle wastage; enhancing joint mobility; keeping the heart healthy and maintaining good blood flow; maintaining general organ condition; reducing fat and avoiding obesity; preventing boredom and reducing tiredness. It isn't normally necessary to suddenly stop all of the games and forms of activity that your dog enjoys and move to a sedate, 'senior' programme; if your dog is still up to it, simply reduce the intensity of the exercise as required, adjusting it to suit changing needs and abilities.

Some golden rules for exercising a senior dog

- *Talk to your vet. Make sure that the exercise regime you have in mind is appropriate for your pet's age and condition*
- *Start with a warm-up. Before you undertake any fast or strenuous walking, stroll around with your dog in a leisurely fashion for a few minutes to get everything moving. This will enable your dog to perform better, and reduce the chances of a muscle injury. Sometimes, a little massage is beneficial before and after exercise – follow the techniques in a good canine massage manual (see* The Complete Dog Massage Manual, *published by Hubble and Hattie) if you are going to do this*
- *Exercise little and often. Several short walks daily – not just at weekends – are better than one long daily route march. Walking this way will help to reduce strain on your dog's bones and body organs*
- *Provide water. Many senior dogs drink lots of water due to kidney problems, but even if this is not an issue, you should carry a bottle of water to prevent dehydration, as well as a collapsible water bowl*
- *Be guided by your dog; if he shows signs of fatigue, such as lagging behind, lying down for a rest or panting excessively, he has probably had enough*
- *Where and when you walk are important. Try to include a good proportion of footpaths and fields; these are kinder to joints than hard surfaces like pavements. Keep in the shade as much as possible on hot days. Don't always walk your dog in the same direction around a field: alternate by going in the other direction. Avoid walking too far in very cold, very hot or very wet weather. A waterproof dog coat will help to keep his body warm in winter. When returning from a wet walk, always dry your senior dog's coat well with a good towelling rub*

Lying down in the middle of a walk is a sure sign that your dog is tiring. Allow him to rest before slowly continuing home, stopping for any more breaks that he may need.

Senior workouts

Depending on the fitness of your pet, there are all sorts of exercises that can help to keep him healthy and supple. The least strenuous of these need be little more than a gentle, ten- or fifteen-minute stroll, so that toilet duties can be attended to; but even this will help to keep muscles and joints supple, as well as provide mental interest and stimulation. And of course, your canine friend will derive pleasures unknown to us by having the opportunity to sniff out the 'p-mail' and to leave his own scent marks. If your dog is up to it, you can extend the duration of this walkabout, and even up the pace to make the experience more aerobic. Just keep an eye on your dog and watch for any signs of fatigue. A more active senior dog that has no arthritic or similar issues can be encouraged to be more aerobic if you walk at a brisk pace, or even jog. Seeing a dog trotting effortlessly beside us as we puff along can

These two senior dogs enjoying their swim are exercising their muscles without putting a strain on their joints.

often be a reminder of how well adapted dogs are for running – and how unfit we are!

Sandy beaches are an ideal place to exercise the senior dog. The sand is a great surface on which to trot and run, and a little gentle swimming in the sea is also very good exercise for your dog. Paddling while supported by the water helps to exercise the muscles and associated tissues without putting undue strain on the rest of the body – a kind of hydrotherapy, in fact. Beware, however, that strong currents may be present, so don't let your dog swim out too far. Regular exercise on the beach also helps to keep the claws trimmed, due to the abrasive nature of the sand. You can make the experience more interesting for your dog by throwing a ball or something similar for him to chase. In contrast, rocky or stony beaches are not good places to take your four-legged friend; even an agile, younger dog can easily slip on wet rocks and stones, and if this should happen to an older dog, he may suffer serious injury as a result.

Instead of taking your dog to the beach, you

(Pages 53-55) Agility exercises involve getting your dog to go through a kind of canine obstacle course, such as running through tunnels, clambering up and over ramps, and clearing 'fences.' Don't let your senior dog attempt to tackle obstacles that are beyond his level of fitness or capability, however.

may be able to arrange for him to enjoy a dip in a special canine swimming pool. Choose a facility that has knowledgeable, properly qualified staff who can design a programme of exercise tailored for your senior's needs. Check out the internet or look through your trade telephone directory to find out where you can take your dog swimming. (There is more about this treatment in the chapter entitled *Special care*.)

And just because your dog is now a senior, it doesn't mean that there will be an instant cessation of all the games he usually plays. However, the older dog may tire more quickly now, meaning games will be of shorter duration, and he may sometimes need a little encouragement to get interested in the first place. Your dog will probably dictate the pace and frequency of games, and let you know when he has had

enough for the moment, but it's also up to you to make sure you keep an eye on proceedings, and don't let him get tired or over-exerted.

Agility games are another good method of exercising a dog, while at the same time keeping him interested in what he is doing. Agility exercises are intended for dogs of all ages, and can be very beneficial for some older dogs, but if you want to involve your senior in this kind of activity, make sure your vet is consulted, in case it isn't appropriate. Agility games involve human handlers guiding dogs over beams, through tunnels and obstacle courses, around poles, and so on. There are organisations and clubs which promote these activities. If you don't want to join these, however, it may be possible to watch some of them in action, and then devise a simple agility course in your garden. If you decide to build your own course, do ensure nothing is constructed in such a way that it could injure your dog. Agility games help coordination and balance, as well as keeping the body in trim. For lots of ideas on games of various levels, see *Dog Games – stimulating play to entertain your dog and you*, published by Hubble and Hattie.

Mind games

One of the key ways of maintaining fitness and health is to prevent an older dog from becoming bored and listless. Dogs are naturally inquisitive and playful, and even if yours is no longer especially active, you can boost his interest in life, and encourage a little exercise at the same time. For example, you can devise games that involve your dog searching for treats (don't use fatty treats; tiny pieces of carrot are much better). Simply hide the treat under the edge of a blanket and encourage your dog to sniff it out. When this becomes too easy, hide the treat in more obscure places, and praise your dog lavishly when he finds it. Doing this for five or ten minutes, once or twice a day, will also give your dog something to look forward to. For a more sophisticated version of this game, you can buy a specially designed plastic ball that contains holes, inside which you place small treats. By nosing the ball around, sniffing out the treats inside, eventually your dog will cause the treats to drop out through the holes. Other products on the market include trays with covers that must be slid open by the dog in order to get the treat. If your dog likes to play chasing games, try throwing a toy and then hiding, so that he has to come and seek you out to bring it back.

This Collie is clearly enjoying hunting for the treat that is hidden beneath one of the sliding covers. Note that the tray is placed on a thick mat so that it doesn't move about.

Pre- and post-exercise massage

As well as using massage to warm your dog's muscles prior to exercise, you can apply gentle massage to help the cooling down process afterwards. Before exercise, massage will help to encourage blood flow to the muscles and tissues, improving their efficiency and helping to reduce the chance of injury. You can begin pre-exercise massage about 20 minutes before any activity. After exercise, massage will assist in ridding the body of toxic products present in the muscles and other tissues, and can also help speed the repair of any slight tissue damage that may have occurred, as well as calm and relax your canine friend. This can be very beneficial for an older dog who may be more prone to such problems than younger animals. A gentle stroking technique known as effleurage can be applied for pre- and post-exercise massage. As mentioned previously, however, do not attempt this until you have been shown the correct way to do it, either by following instructions in a book intended for this purpose, such as *The Complete Dog Massage Manual*, published by Hubble and Hattie, or by having the technique demonstrated to you by an expert. Speak to your vet before using massage, especially if your dog has any kind of condition that may mean it could be detrimental to his health.

This dog is wearing a specially designed harness for use in the car. Merely holding onto your dog, or sitting with him on your lap, is no substitute for fitting a proper safety device like this one.

Car travel

In the same way that you will make adjustments to his routine at home, you will also need to adapt and adjust conditions in the car so that your senior dog is comfortable on journeys.

Unless you have recently acquired an older dog, the chances are that your companion will be accustomed to riding in the car, having grown up with the experience. Whether or not your canine passenger is a seasoned traveller, it's important to ensure that he rides in safety. It is remarkable how often one sees a car with its human passengers safely strapped into their seats, with their dog sitting on the seat behind or beside them, or loose in the area behind the back seats. This is a highly dangerous state of affairs, however, because if the car is involved in an accident, or if the driver needs to brake suddenly, the dog could be thrown about, potentially causing great injury to himself, or other passengers.

There are several ways to restrain your dog comfortably and safely in a car, and while advice about safe restraint applies to all dogs, it is especially essential that an older, possibly frail dog is protected in this way. If you have an estate car or a hatchback, you can buy a safety grille or grid that fits behind the rear seats; this will prevent the dog from being projected forward if the car brakes suddenly. Many designs have universal fittings that allow for plenty of adjustment, so they can be transferred from one similar-sized car to another, if necessary. Most grilles are designed with an open, latticelike arrangement so that the dog doesn't feel cut off from his owner. A similar arrangement to the metal grille just described comprises a strong, nylon safety net to separate the passenger area from the boot area. These are inexpensive to buy, and they can normally be attached to structures like the passenger grab handles. Such devices also have the added benefit of preventing any other objects, such as cases or boxes, from being projected forward during sudden braking.

Another very practical solution is the dog travel cage or crate. Like the grilles just mentioned, these also come in various sizes, normally with an open-mesh design to avoid the dog feeling 'hemmed in,' and a solid base onto which a bed, a rug, or newspapers can be placed. Some have several doors, as well as an escape hatch. They can even be supplied with inner partitions. To ensure they can be accommodated in small hatchbacks, some designs are tapered on one side to allow for the sloping shape of the car's rear end. If your dog is going to travel in the car on one of the passenger seats, then the best solution is to fit a safety harness. This generally consists of adjustable padded straps that are secured around the dog's shoulders and chest, using quick-release buckles. The harness is then attached to one of the seat belt mechanisms in the car. In the event of sudden braking or inertia, your dog will be safely restrained. Some dogs may take a while to get used to the harness around their body, particularly if they have been accustomed to having more freedom in the car, but an older, less active dog will often accept the arrangement more readily, especially if a human passenger is occupying one of the other seats and can offer plenty of reassurance and attention on the journey!

Travel equipment

As well as ensuring your dog's safety, his comfort should also be considered. A car journey, particularly a long one, can sometimes be an endurance test for an older dog, so anything you can do to make it more pleasant for him will be appreciated. Your dog may prefer to lie on his favourite rug, on a specially designed dog cushion, or in his bed if there is room for it – even if he is travelling on the back seat. It goes without saying that a dog travelling in the boot area of a hatchback should have plenty of soft bedding, so he doesn't become bruised by anything when the car moves about. You can also buy disposable floor mats for dogs, which can be very useful if yours has a weak bladder, and can also help to mop up water spilled from water bowls. Speaking of water bowls, it's all too easy for a dog to tip these up when travelling; many dogs seem to have a knack of standing in them. To help prevent spillage, use a water bowl with an overlapping rim.

This plastic water bowl has an overlapping rim that helps to prevent water from spilling out when travelling. The rim can usually be pulled off when not required. Some dogs take a little while to get accustomed to drinking from a bowl with a rim, but soon get the hang of it.

Another piece of equipment you might like to consider is a waterproof seat cover. These clip over the back of the seat, and prevent the fabric becoming soiled by any accidents.

Young dogs are usually very eager to leap into the car – the event signifies something exciting such as a walk or a holiday, and they don't want to be left behind. However, as your dog gets older, he may experience more difficulty in getting into the car, particularly if he needs to make a leap up into the boot area. He may also begin to find it hard and uncomfortable to jump out of the car.

This specially designed extending ramp enables a senior dog to access a car with ease. Note the non-slip surface that helps with grip.

Beginning to find these actions difficult can cause anxiety for your dog, since he doesn't want to be left behind. If your dog is relatively small, you can, of course, lift him into and out of the car quite easily, but this isn't really always a practical proposition for a dog the size of a Labrador Retriever or larger, and even if you can lift your dog, it may be uncomfortable for him. The simple solution here is to invest in a dog ramp or steps, obtainable from many pet suppliers, and make access into and out of the car much safer and easier. A typical ramp for a medium-sized dog is about 59in (150cm) long and about 19in (48cm) wide, and many designs are hinged, so they can be folded to take up less space when not being used. If you are handy at DIY, it is possible to make your own car ramp., but if you do, make sure you cover it in a material that prevents your dog from slipping, and also ensure that one end fits securely to the car.

Some of the other tips for travelling with the older dog apply just as readily to dogs of any age. Don't forget to carry a towel, some food, a bottle of water for replenishing the drinking bowl,

When you break a car journey, give your dog a chance to have a walk about and then stretch out in comfort for a while.

and some bags to collect the waste when your dog relieves himself (and be prepared to make more frequent comfort breaks with your senior passenger on board). Make sure, too, that your dog is never sitting directly in the heat of the sun coming through a window when travelling; if necessary, move him to the other side of the car, or fit a sunshield to the window. Provide plenty of ventilation, but do not allow your dog to put his head out of the window, even if he is restrained. You can fit a window guard that prevents your dog from doing this, while allowing fresh air into the car. And never leave any dog alone in the car, whatever his age and whatever the weather. Apart from the risk of the dog being stolen, the temperature inside a car can rise to lethal levels very quickly – even with the windows open – causing the animal to suffer an agonising death. In some countries, it is, in fact, illegal to leave a dog in the car under any circumstances, and a heavy fine awaits those who fall foul of this law.

A canine travel pack, with everything needed for the journey.

Holidays

Dogs are always a particularly important consideration around holiday time, whether you take them with you, or have someone care for them whilst you are away. Although dogs are extremely adaptable animals by nature, and will normally fit happily into your lifestyle, they can often find the break in their daily routine around holiday time a stressful experience. Older dogs, especially, can be unsettled by the unfamiliar surroundings of a holiday destination, or by being separated from their owners for perhaps several weeks, if they are not able to join them.

If it's feasible, taking your pet away with you is often the best option – and probably more fun for you, as well as for the dog. This will prevent any pining on the part of the dog – and sometimes on the part of the owners, too! The best holiday

The introduction of the so-called Pet Passport Scheme, or Pet Travel Scheme, means that dogs can now travel from the UK to other parts of Europe, without the need to undergo a period of quarantine on their return. However, your pet will need to be vaccinated against rabies – a deadly disease that is rife in some parts of Europe – and will also need to be microchipped if this has not already been done. Despite this easing of the laws concerning pet travel, one must consider very carefully the decision to take an elderly dog abroad. The extended travel necessary to reach your destination – probably much of it spent in the car or on board a ferry – as well as the likelihood that you may be visiting a country with much higher temperatures than you experience at home, may mean that you decide to leave your dog at home.

destination is somewhere like a rented cottage or a caravan that allows dogs. You may find, however, that accommodation of this type is sometimes of an inferior standard to that which is advertised as 'no pets allowed.' Even when accommodation allows pets, it is worth checking the 'dog-friendliness' of the surroundings. You don't want to be walking great distances before your dog can be exercised or allowed to relieve himself, especially in the case of an old dog. If your dog has a tendency to chase other animals, also check that the accommodation isn't surrounded by fields of sheep or cows. It may also be worth enquiring about any other potential problems you can think of (such as the possibility of lots of steps to negotiate, proximity of busy roads, and so on).

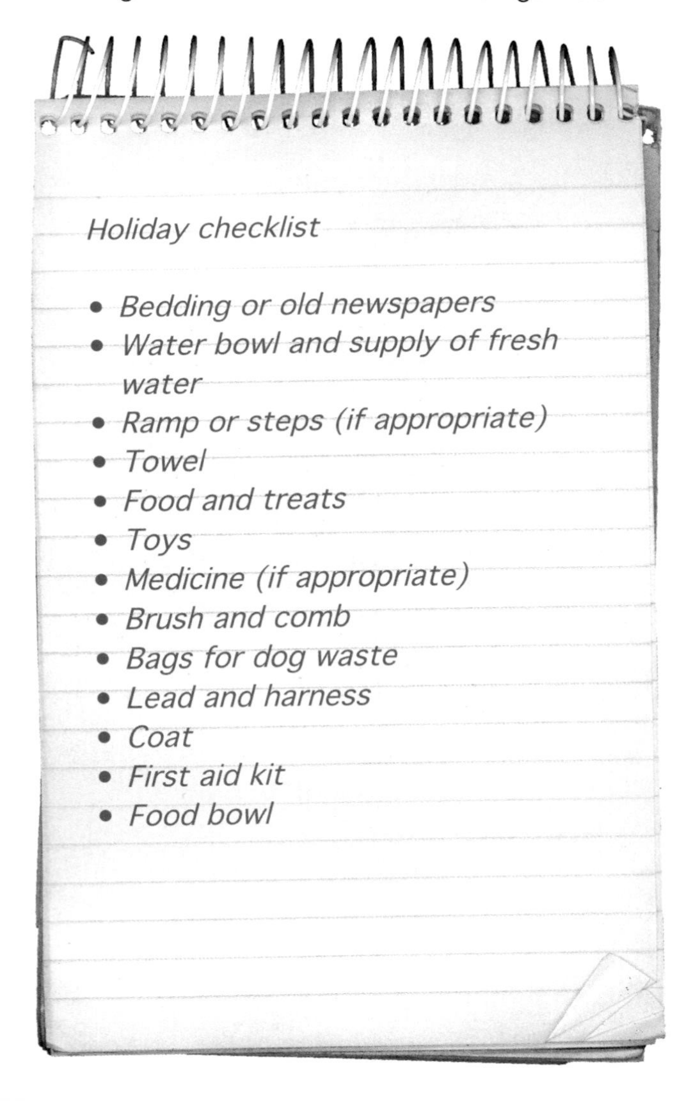

Many hotels allow well-behaved dogs to stay, but there is normally a surcharge for this. Before booking, it is worth enquiring whether your dog will be allowed to stay in the room for periods when you go out. If not, you must be prepared to have the dog with you at all times. This is an important consideration, because an older dog may get very fatigued plodding around all day if you are out and about, and will considerably restrict your ability to visit restaurants and many other places. Bear in mind, too, that many beaches may have a ban on dogs, especially during the summer months.

Some holiday companies actually cater for people who want to take their dog with them on holiday, and offer plenty of facilities to make life easier and more enjoyable for all concerned. These companies frequently advertise in newspapers and on the internet.

Another good option for an elderly dog is a boating holiday on a canal or river. Many boat hire companies allow dogs on board their vessels, and your pet can sit comfortably in the cockpit, watching the world go by while the boat cruises along. Exercise can be taken at a leisurely pace when you moor up, and it is quite likely that you will find plenty of riverside pubs where your dog will be welcome, too. Keep a lifejacket on your dog when the boat is travelling, and make sure your pet is safely under control at all times, especially when mooring up or when going through locks.

Try to maintain your dog's daily routine as much as possible, even when you are on holiday, by providing exercise and meals at the usual times. Many of us, if we are reasonably fit, use our holiday as a chance to get some additional exercise, but you shouldn't subject your elderly dog to more exercise than he is accustomed to; this will simply tire him out and could exacerbate or bring on muscular or other problems. Certainly don't make your dog endure extended walks if he is suffering from hip and other joint problems. It's worth discussing your walking plans with your vet before you go away, in any case, so that he or she can advise about what is suitable for your dog.

Boarding arrangements

If you decide not to take your companion on holiday, there are several other options to consider. Many people choose to place their dog in a boarding kennel, and if you do this, use one that has been recommended by someone you know. If you have no recommendations, look in your local telephone directory to find an establishment or two near you. In any event, you should arrange to visit the kennels well before you need to go away, so that you can ensure you are satisfied with the facilities on offer. Remember, too, that good establishments get very busy around traditional holiday periods, so it pays to choose and then book (if satisfied) well ahead. You will need to ensure that your dog's vaccinations are up to date. Some kennels also insist on proof of kennel cough treatment prior to accepting a dog for boarding. If your elderly dog has any special needs or is undergoing any treatment regimes (such as tablets on a regular basis), remember to make this very clear to the kennels before you book, in case it is unable to fulfil these requirements. In any event, a written list outlining your pet's needs should be given to the kennel well in advance of your dog's stay there, and also presented to the member of staff who takes the dog in. You should also supply some contact details, including those of a relative or friend, and the vet that treats your dog. Your pet will also appreciate having his favourite blanket and a few of his toys to remind him of home.

Although your elderly dog will undoubtedly be safe and sound in kennels, owners often prefer a more 'personalised' type of care for their pet – especially for an older dog. For example, it may be possible to arrange a house-sitting service, whereby your pet is looked after in his own home. This usually involves an experienced, often retired person staying in your home for the duration of your holiday, which means that your dog can remain in familiar surroundings and stick to his daily routine. As before, make sure that there are no problems concerning any special care your dog might need, and ensure that the carer has details of how to look after your dog, including diet details, when he has treats, walks, etc.

Another popular scheme exists whereby experienced dog-carers take in one or several dogs to live with them in their own home while you are away. Always check out the insurance, credentials and references of such services before you book. Also, bear in mind that they may have limited capacity in terms of the number of dogs they can take in at any one time, so don't leave it too late. You will find that such services are inevitably more expensive than regular kennels, but on the plus side, your four-legged friend will probably receive more attention than he would do in kennels.

FEEDING THE OLDER DOG

A toned-looking eight-year-old Springer Spaniel cross. Although neutered, a sensible diet and plenty of exercise have maintained this senior dog's weight at the right level.

One of the key factors contributing to dogs living significantly longer today is a much better understanding of their nutritional needs. In addition to general dog foods, there are now special life-stage diets that cater for dogs at both ends of the age spectrum. Foods for older dogs, or seniors, are widely available, and it is usually recommended that shorter-lived breeds of dogs be switched to these diets when they reach the age of about six, and that most other dogs are introduced to them at around seven years of age.

The diets are formulated to reflect the changing nutritional requirements of dogs at this stage in their lives, and they should not be confused with prescription diets available from your vet for conditions such as kidney (renal) failure (see page 70). Senior diets contain easily digested protein, combined with a range of ingredients – such as Vitamin E – known to help boost the immune system. Such foods also have significantly fewer calories than standard dog food, because older dogs are more likely to gain weight as they age. There can be several reasons why this happens. Firstly, because neutering affects the metabolism, a dog will put on weight after surgery of this type unless her food intake is reduced. Dogs will also tend to be less active as they grow older, particularly if afflicted by any joint disorders, and if they continue to eat the same amount of calories, this is likely to result in unwanted weight gain, which can mark the start of a downward spiral, whereby having gained weight, a dog then continues to become less and less active, causing her weight to increase. Piling on the pounds will, in turn, increase the strain on ageing joints so that exercise becomes more painful, further exacerbating the problem.

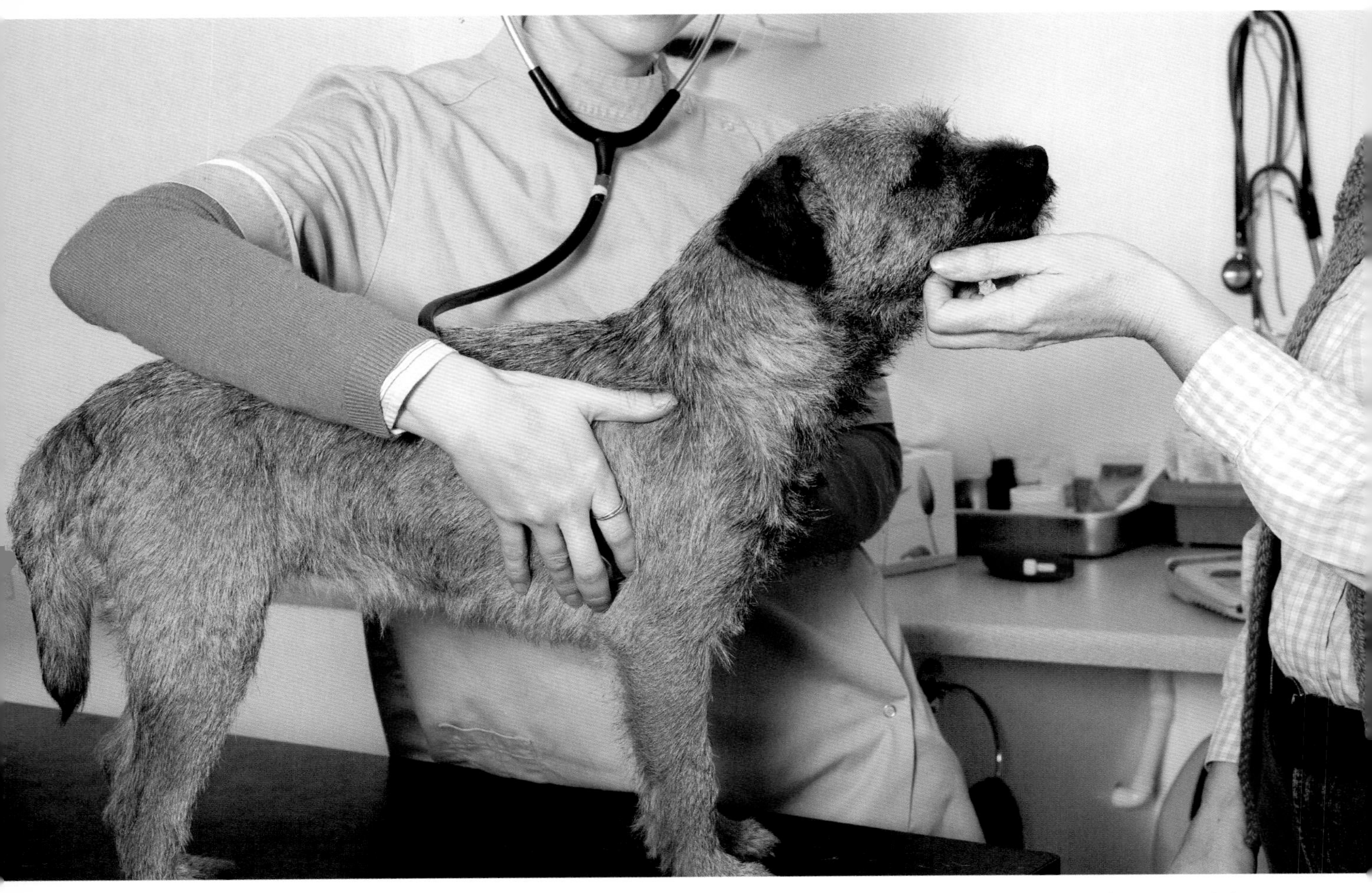

A vet checks the ribs of a Border Terrier. If the ribs cannot be felt at all, then your dog is seriously overweight.

Taking action

It's not just a question of adjusting your pet's diet as she grows older, therefore, but also of monitoring her weight carefully. This can be done very easily using bathroom scales. If you have a small dog, persuade your pet to sit on the scales and then simply note her weight. However, with a larger dog, you may need to lift her up and weigh both of you together, before deducting your own weight from the total. Always weigh your dog first thing in the morning, before providing any food, and keep a note of the figure somewhere readily accessible, such as in a diary. Do this every two weeks, and you can then build up an accurate picture of whether or not your pet is gaining weight.

It is much easier to act before your canine

These veterinary scales will give a very accurate readout of a dog's weight.

A range of special foods is available for dogs, which can cater for all kinds of problems, from obesity to deficiency problems. Your vet will be able to advise you.

friend becomes seriously overweight. The average weights for purebred dogs are usually included as part of the breed standard, so they are easy to find; in the larger breeds, dogs tend to be slightly bigger and heavier than bitches. Even with mongrels, however, there are still indicators that will reveal whether or not your dog is overweight, the best of which are the ribs. These should be just discernible under the skin when you run your hand along the side of your dog's body (although this is harder to do if your dog has a long coat).

If you suspect that your dog is too heavy, contact your vet for advice. Many practices run special weight-loss clinics for pets these days, and can provide a special weight-loss plan that will be individually tailored to your dog's needs. Sticking to this will require some willpower on your part, and that of other family members – especially if your dog has been accustomed to receiving tidbits on a regular basis, since these will need to be cut out. It's important to bear in mind that allowing your pet to remain overweight will worsen underlying health conditions such as heart disease, potentially shortening her life as a result. Furthermore, an overweight dog will undoubtedly suffer a reduction in her quality of life as well.

If your dog has become seriously overweight, it may well be that a typical 'senior' food is not really sufficient to deal with the problem. Under the circumstances, your vet may advise switching to a special obesity diet that will help her to lose weight more effectively. There's also now a weight-loss pill for dogs available on prescription from a vet that may be recommended in certain cases.

Kidney concerns

As dogs grow older, the ability of their kidneys to concentrate urine is likely to be reduced. The use of a special renal diet may be recommended by your vet, which will help to alleviate the worst effects of this problem. A special renal diet will contain a higher level of protein than normal, helping to contain the loss of condition that is linked with this degenerative ailment. A food of this type will contain supplements of the so-called 'water soluble' vitamins – members of the Vitamin B group (important in terms of the body's metabolic functions), as well as Vitamin C. This is important, since these vital vitamins will be lost from the body in greater quantities than normal, due to the increased volume of urine being produced, and thus, there is a risk of a deficiency arising.

The mineral balance is also affected in cases of chronic kidney failure, with the body's ability to absorb calcium being compromised as a consequence. A shortage of this vital component of bone is dangerous, particularly over a period of time, and can lead to a weakening of the skeleton and a dangerous increase in phosphorus levels in the blood as well. Renal diets are supplemented with extra calcium, and have lower levels of phosphorus than normal to help counteract these problems. If you have always cooked your dog's food yourself, seek advice from the vet as your pet becomes older; additional supplements of vitamins and calcium may now be advisable. If you are using a prepared 'senior' diet or a prescription food, it is unlikely that you will need to use additional supplements, since such foods are carefully formulated.

You may find that other dietary changes are beneficial. Although dry food can be helpful as far as a dog's dental care is concerned (see *Diseases and the older dog*, page 82), it is often less palatable than canned food or pouches, which are often described as 'wet food' because of their higher water content. This can be an important consideration, since chronic renal failure can depress the appetite.

Older individuals are also at greater risk of suffering from constipation. This can be prevented, to a large extent, by mixing a teaspoonful (5ml) of liquid paraffin (available from pharmacies) with your dog's food, if necessary. Alternatively, regularly adding a handful of bran to her meal should help with this. Flatulence can also become more of a problem.

Offering your pet special charcoal biscuits may bring some relief for flatulence, or alternatively, there are herbal treatments for dogs, such as garlic tablets, that can also help.

Dried 'complete' foods are specially formulated to contain all the components your dog needs for a healthy diet, though can tend to be a little dull, even though available in various flavours.

Chewing dental sticks daily helps to clean the teeth.

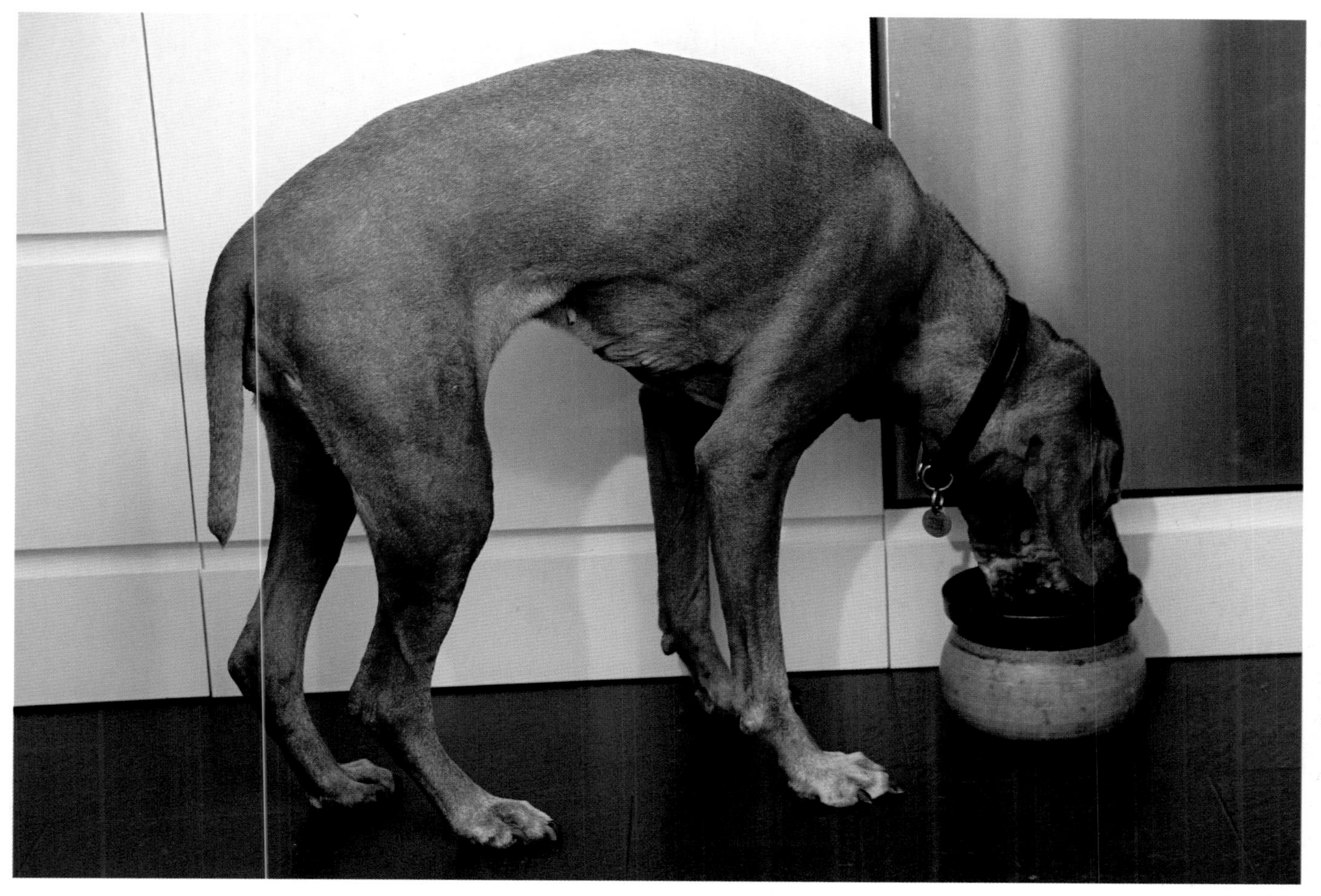

An old dog, especially a large breed, may struggle to reach her bowl comfortably. In such cases, the bowl can be raised up, as shown here ...

Joint supplements

As dogs grow older, they are more liable to develop joint stiffness, often linked to 'wear and tear' on the joints, with the effects being most pronounced in dogs afflicted by hip dysplasia and those that are overweight. The impact of degenerative disease of this type, described as osteoarthritis, can be reduced by slimming your dog, if she weighs more than she should, although supplements can help in many cases. The key component in supplements of this type is glucosamine, often in combination with other components, such as chrondroitin. Glucosamine helps to repair the damaged cartilage within the joint, while chondroitin helps to prevent further damage and ensure a buildup of fluid in the joint

... you can also buy an adjustable stand that makes it possible to set the bowl at exactly the right height for individual dogs to eat or drink comfortably.

space. This combination can improve overall functioning of the joint, giving greater flexibility and making it less painful for the dog to move. Another ingredient called MSM may also be incorporated in products of this type, reducing inflammation and improving joint mobility.

Ask your vet's advice about using a joint supplement to help improve your dog's condition. Always use a product that has been formulated especially for dogs, rather than for people, or for other animals, such as horses. It will probably take between four to six weeks to see any improvement, so do not expect an instant result. If it proves effective, it can significantly improve the quality of your dog's life, and may also mean that you can reduce or dispense with painkillers for your pet's painful joints. Your vet may actually recommend a course of injections that provide a more immediate effect at the outset, and then the use of a suitable supplement to maintain the wellbeing of the joints.

Food allergies

These can develop at any stage in life and are not uncommon in older dogs, but can be difficult to diagnose. The effects can also be varied, resulting in itchy skin, a rash, or even a digestive disturbance. Some breeds, such as Dobermanns, may be more susceptible to this type of problem than others. It is also important to distinguish, in the first instance, whether your dog has a food allergy that affects the skin, or an environmental allergy, that may be linked, for example, with a washing powder used to clean her bedding.

You will need to adjust your dog's diet, removing the source of protein that she was eating. For example, you may need to eliminate beef, and replace it with another protein source – like duck or venison – which she has not eaten before. If your dog's condition begins to improve after a period of about three months, this increases the likelihood that she was suffering from a food allergy. It's not just the meat component of her meal that can be an issue, however, as it could be the carbohydrate component, in the form of wheat, for example. In this case, replacement with rice or sometimes barley may be needed. But since each potential allergenic component needs to be substituted in turn, tracking down the cause can be a very slow process.

Testing in this way once entailed having to cook a specialised diet for your dog, but it is now possible to purchase specially formulated and balanced hypoallergenic exclusion diets. These should help to pinpoint the allergen responsible for the condition more easily, as well as aiding your dog's recovery. When investigating a suspected food allergy, it's also very important to prevent your dog from eating any other food, because even just a few tidbits, for example, may interfere with the results.

Liver diets

Just like the kidneys, a dog's liver can begin to fail, particularly as she grows older. The liver is a key organ in the body, being involved in the metabolism of food, as well as breaking down drugs and toxins. It's also concerned with maintaining the hormonal balance in the body, as well as the blood and clotting mechanism.

Careful dietary adjustments mean that the health of the liver can be maintained. Basically, reducing the protein level in the diet means that the liver will not be exposed to such high levels of ammonia, which it would normally break down and neutralise in the blood. In cases of unchecked liver failure, ammonia is likely to build up in the blood. This ultimately affects the dog's behaviour, producing signs such as profuse salivation, tremors and, most characteristically, head-pressing, where the dog leans her head against any available objects, such as chairs.

Diets to counter liver disease are also likely to contain complex carbohydrates. These take longer to digest than simple sugars, which means that the absorption of glucose from the intestines to the liver takes place over a longer period than usual, so the liver does not need to work so hard in processing it. A zinc supplement is also incorporated into food of this type, because it has an important part to play in the breakdown of ammonia in the liver. The addition of extra fibre helps to remove ammonia before it can be absorbed into the body, too. Antioxidants such as Vitamins C and E, as well as taurine, that help to protect the liver's ability to function and hopefully recover in due course, also feature in specially formulated foods of this type.

USING YOUR VET

It is vital that vaccinations are maintained, otherwise a chance encounter with an infected dog in a park, on a beach, or in any other public place could carry a real risk of disease being transmitted.

Regular health checks will become increasingly important as your dog grows older. Be guided by your vet, but, on average, perhaps two or three visits over the course of the year may be a good idea. More frequent trips might be necessary, however, if your dog is suffering from a condition that requires regular medication, so that a check can be kept on his progress. In some practices, you may not see the same vet on each occasion, though should be able to arrange appointments so that you do see the same vet each time; this can be advantageous as then it's much easier for a more consistent assessment of your dog's condition to be made. This is especially helpful where ongoing treatment is concerned.

Vaccinations and check-ups

Older dogs should continue to have their vaccination boosters, just as they did when they were younger. Indeed, protection may be more important at this stage, simply because his immune

Here, a vet gives a booster injection in the scruff of the neck.

system will probably not be working so effectively now, which can leave an older individual more vulnerable to dangerous infections that can strike at any time. Vaccination safeguards against potentially fatal canine illnesses: distemper, leptospirosis, canine adenovirus and parvovirus can kill dogs at any age – not just puppies. Although these boosters are usually administered annually, it has been suggested (and some vets concur with this view) that yearly boosters are not necessary, and can be done every two or three years, certainly for some of the diesases. Talk with your vet about this and choose the right option for your companion.

If you have habitually taken your dog on holiday with you when he was younger and more energetic, and are now thinking of using boarding kennels instead, you need to be aware of the highly contagious respiratory illness known as 'kennel cough,' or infectious tracheobronchitis. There is no single cause of this illness, though a variety of microbes, notably canine adenovirus type II, and the bacterium known as Bordetella bronchiseptica, can be implicated in this condition, with the incubation period lasting up to ten days. The most obvious symptom is a dry cough, which can be followed by retching, and a loss of appetite.

In many cases, kennel cough will resolve itself after a period of five days or so. But even so, and especially in an older dog, you should seek veterinary advice, otherwise, there is a risk that the infection could develop into bronchopneumonia, which an older dog may not be able to fight so

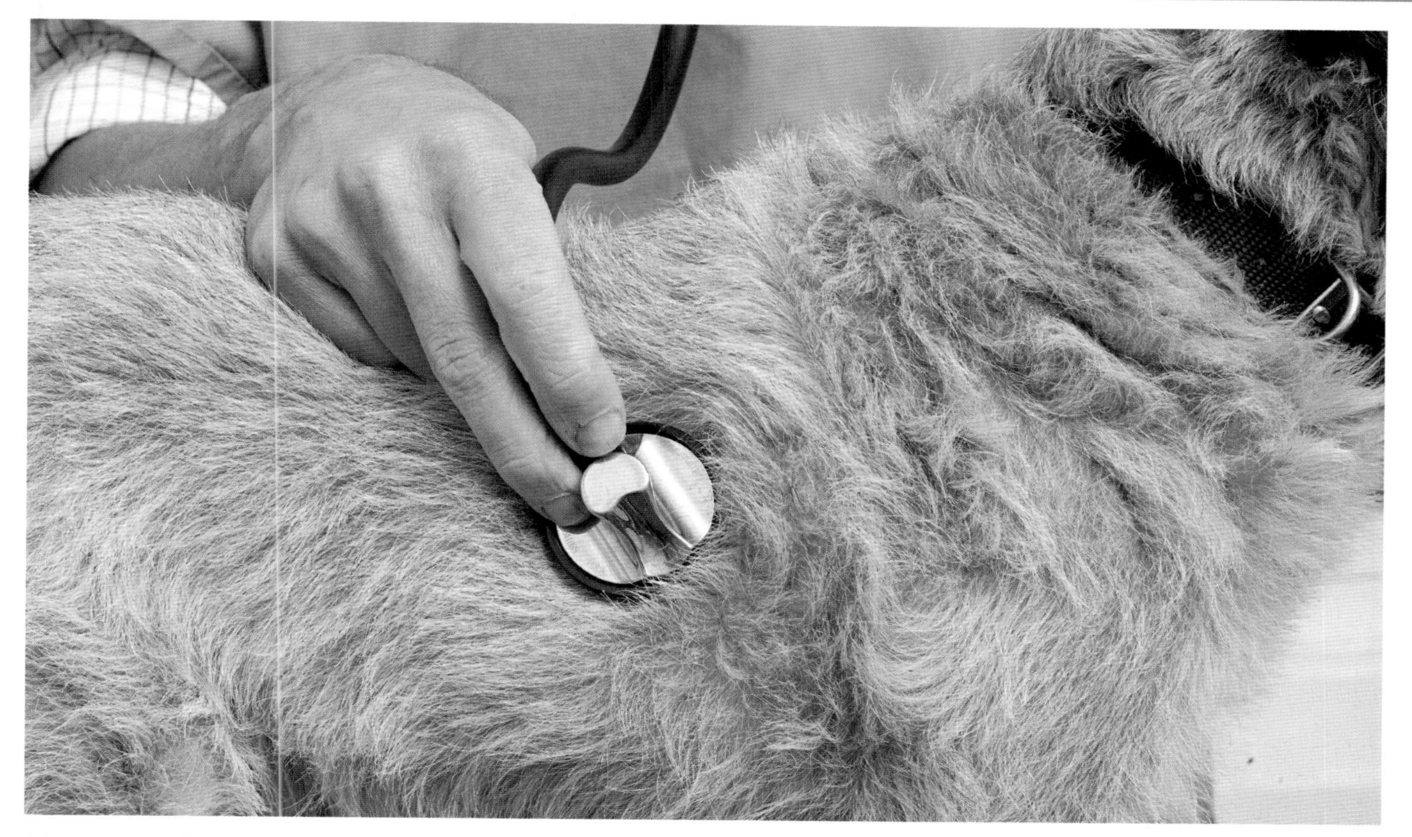

Listening to the heart will help to detect any abnormalities, such as murmurs. Further tests may then be necessary to establish the cause of the problem.

effectively. Symptoms of bronchopneumonia include a high temperature, often accompanied by discharges from the nose. The seriousness of bronchopneumonia emphasises the value of having your pet vaccinated against the common causes of this illness, should you need to use boarding kennels. Even if it does not prove fatal, the infection is very unpleasant for a dog. The vaccine is simply squirted up the nostrils, rather than being injected into the skin.

There is a great deal that a vet can do to monitor the health of your dog. For example, listening to the heart will detect any significant abnormality, and then your vet may recommend a further examination known as an electrocardiogram. Blood tests can be significant in detecting and monitoring a wide range of conditions. A basic urine test can also reveal much; it can detect possible diseases like diabetes mellitus, and can provide an indication of kidney function as well (see also the chapter entitled *Diseases and the older dog*). Once again, follow-up tests can be carried out in the event of any abnormality showing up.

Collecting a urine sample

If one is required, you will probably be asked to collect and bring in the dog's urine sample. Thoroughly wash out the container being used to collect a urine sample before use. Jam jars can be a particular problem if any sugary residue remains within the jar or on the lid. The best solution is to clean them with a brush, and then, if possible, place them in a dishwasher to be sure they are completely clean. It is important that you do not create a false reading!

Obtaining a sample can be difficult, particularly with a bitch, or with an older male dog that may prefer to squat rather than lifting his leg. A shallow plastic container, such as a clean plastic plant saucer of appropriate diameter, or a low-sided dish, is quite effective for collecting the sample. A pair of disposable gloves is also recommended. Watch your dog closely, and then slide the container beneath the hindquarters at the appropriate moment, being careful not to startle her. With a male dog, collecting the sample directly in the jar is the best option, when your pet lifts his leg. The best time to try to collect a sample is in the morning, when you first let your dog out, because then he is likely to have a full bladder. If you are unable to collect a sample first thing, try taking the dog for a walk; male dogs, especially, may choose to scent-mark trees and lamp posts, so it should still be possible to catch some urine.

It is a common misconception that a large volume of urine is required for test purposes – certainly, you do not need a whole jam jar full. In some cases, less than a spoonful is adequate; ask your vet how much is required. Do not forget to label the container with your dog's name and your own, preferably with a waterproof pen, so the sample can be clearly identified once you hand it over to your vet.

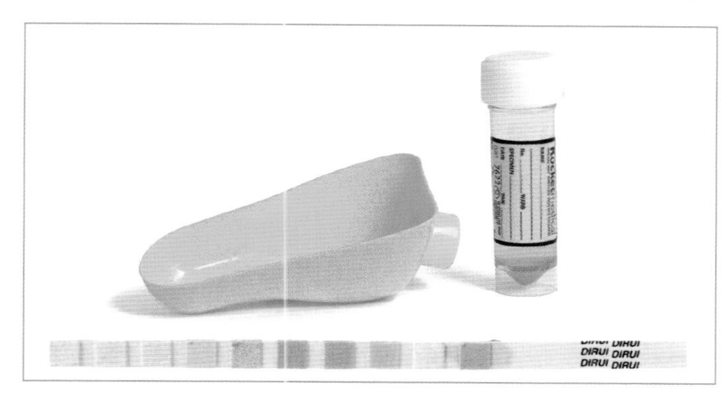

A purpose-made urine collecting scoop and sample bottle. A test strip is dipped into the sample, and the colour it turns is compared with the reference chart to determine factors such as glucose levels in the urine.

Weight watching

As dogs get older, they will become less active. Unfortunately, however, they often maintain their appetite at the same time, and this combination of lack of exercise and a higher-than-required calorie intake will soon result in weight gain. The situation is likely to be even worse in the case of neutered dogs, because this surgery alters their metabolism, and will make them more inclined to put on weight, unless their food intake is cut back, and they are encouraged to exercise regularly. Obesity can be quite insidious in a dog – seeing your pet every day will tend to obscure this condition, particularly if you do not weigh him regularly.

The chapter entitled *Feeding the older dog* describes how to assess whether or not your dog is overweight, as well as including measures to help keep weight at the correct level. It is relatively easy to establish the ideal weight range for most purebred dogs, because this will be specified in the relevant breed standard. In the case of mongrel dogs, be guided by your vet's advice as to whether or not your dog is overweight.

Most practices now have regular weight-loss clinics for dogs and other pets, so take advantage of such opportunities. Start by checking to ensure that your dog has no underlying health problems that could affect a slimming programme, by restricting his ability to exercise. It will then be possible to come up with a weight-reduction plan tailored to your dog's individual needs, which you will be able to put into practice with the confidence of knowing that you are following the right routine. Be patient, though, because it's much harder to help an older dog to lose weight than a younger one. Nevertheless, success can greatly improve the quality of your pet's life, as well as prolonging it, hopefully. Veterinary practices now also offer other services, or can refer you to other professionals who can be of benefit when you are trying to control your pet's weight. One of the most significant treatments is probably hydrotherapy (see *Special care*, page 106). This can be an excellent way for an older dog to exercise and stay fit, even if he is suffering from a disability.

On the other hand, if your dog appears to be losing weight suddenly and unexpectedly, consult your vet, since this could indicate a problem that needs further investigation. It is really important to keep a record of the weight of your dog, perhaps on a fortnightly basis, so that you can detect any weight loss or gain at a relatively early stage.

Your dog may experience impacted anal glands. The discomfort may cause him to 'scoot' (drag his bottom along the ground) or have difficulty defecating. To ease the problem, your vet will be able to manually empty the anal glands.

DISEASES AND THE OLDER DOG

Some breeds are more susceptible to dental disease than others, with small Poodles appearing to be among those especially vulnerable.

As our dogs age, not only do they often become prone to certain types of illness – some of which are more prevalent in certain breeds than in others – but also begin to show physical signs of 'wear and tear,' many of which are not symptomatic of diseases as such, but merely conditions brought about by the advancing years. One of these is greying of the fur around the head and muzzle, which is very easy to see in black-haired dogs. Another is the deposit of plaque or tartar that builds up on teeth.

Dental care

Even if your dog's teeth are brushed regularly, as she gets older, deposits of tartar will usually build up on them. Tartar contains bacteria, and where such deposits come into contact with the gums, they can cause inflammation, sometimes described as gingivitis. Over time, erosion of the gum itself may occur, weakening the teeth concerned, until they become so loose that they may fall out. Typical symptoms include bad breath or halitosis (although this can also be a sign of other disorders,

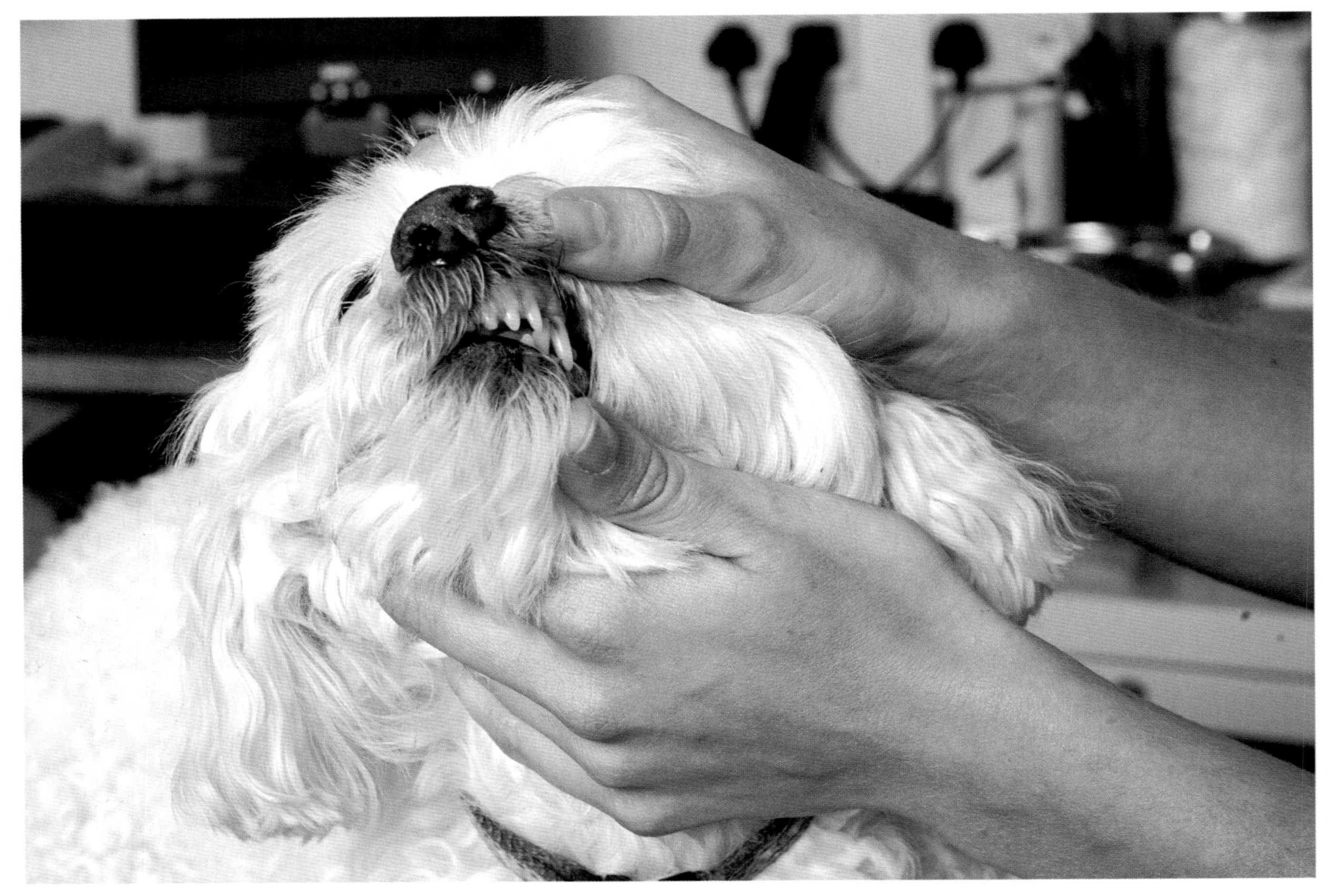

A vet examining the teeth of a dog. A tartar deposit can be seen on the front of the animal's left upper canine.

such as kidney failure), dribbling and sometimes a reluctance to eat. The dog may also rub her mouth with her paws in a bid to relieve the pain and irritation. The area of fur under the lips may show signs of staining and matting, while, in chronic cases, a dog can lose her appetite. As the gum becomes inflamed, it becomes easier for bacteria to gain access to the roots of the teeth, resulting in the formation of abscesses.

The best preventative remedy is to have the teeth cleaned and de-scaled to remove any deposits of tartar that have built up. A vet will normally carry this out under sedation or even a full anaesthetic, because many dogs will naturally object to this treatment. Be aware, though, that anaesthesia always carries a risk, which is especially relevant with an older dog. This is something that your vet will take into account.

A vet may use a magnifying instrument called an auroscope to check the external ear canal for signs of mites.

Ear problems

Chronic ear infections are most likely to strike breeds with pendulous ears, such as Spaniels, who also tend to have a heavy covering of hair as well, extending down into the external ear canal. Shaking the head is a typical sign of inflammation in this case, which is known as otitis externa. Under normal circumstances, protective wax is regularly produced by the cells in this part of the ear, which then dries up and falls out of the ear. This is less likely to occur in the case of Spaniels and similar breeds, however, because the hair in the ear canal causes a blockage, and the earflap itself encourages the moist conditions ideal for an infection to develop. Effective treatment can be difficult, especially when anatomical features favour development of the infection, although trimming the hair may help prevent a recurrence. Certain dogs, such as Dachshunds, are prone to producing excessive quantities of wax as well, which worsens the situation. Both bacteria and yeasts are typically implicated in this sort of infection, and ear mites may also be responsible.

Prompt veterinary treatment will be required to find the cause of the problem. If left, serious complications can result, because the infection may spread into the inner part of the ear, while externally, the irritation may be so intense that the dog injures her earflap. This then swells up as a haematoma forms, resulting from the rupture of blood vessels in the ear.

Should your dog have suffered from repeated

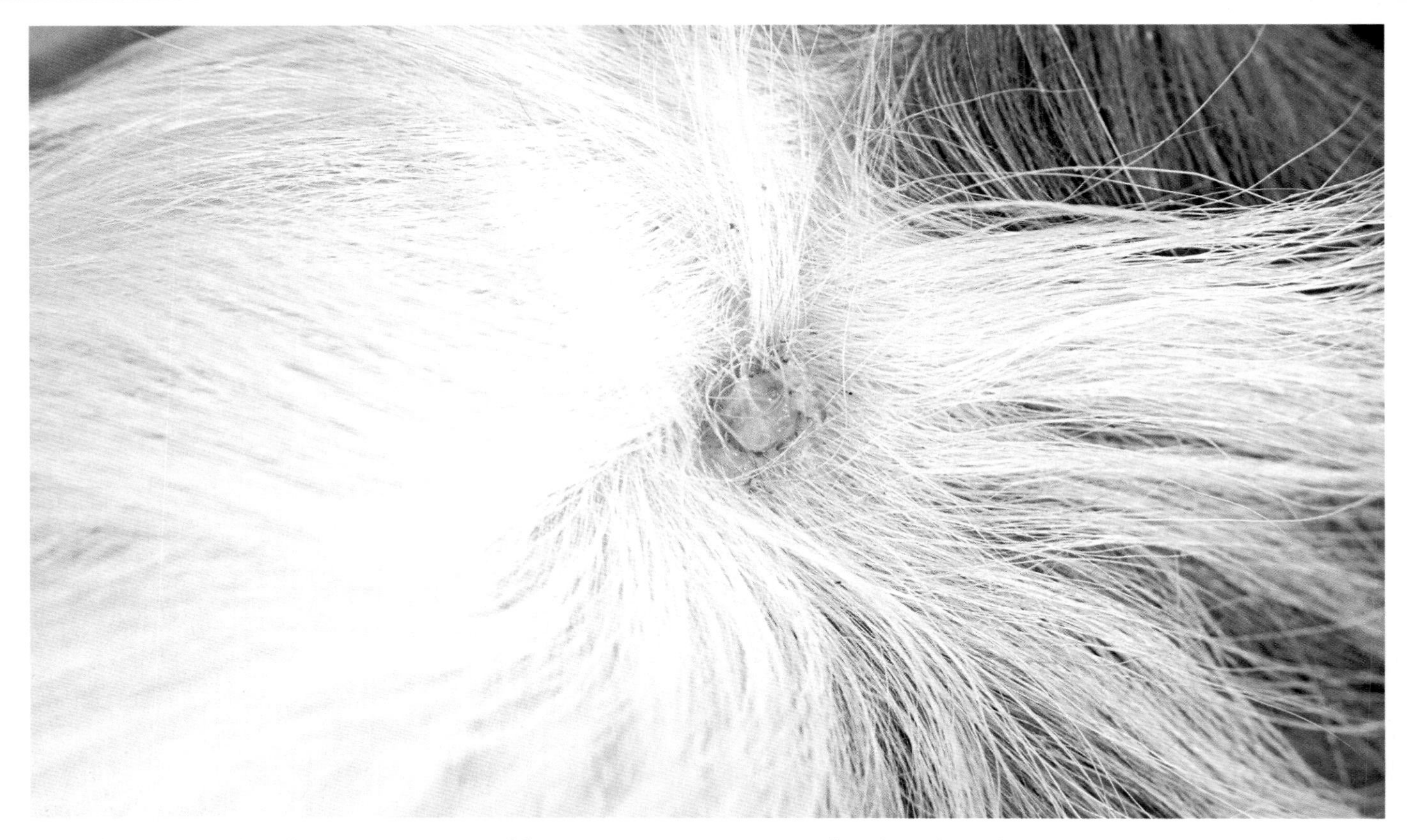

A fleshy wart on the skin. It is quite possible to remove warts surgically, though, unfortunately, they do tend to recur.

infections of this type early on in life, it is quite common for the ear canal to become thickened in old age, as the result of chronic inflammation. Its diameter may then become seriously reduced, to the extent that drainage is impossible. Surgery is then likely to become the only treatment option available, with the external ear canal being opened up to remove the focus of the infection. Known as aural resection, this will resolve the problem, although the operation causes slight disfigurement. With an older dog especially, however, the general health benefits outweigh the cosmetic considerations.

Lumps and bumps

As dogs grow older, they are far more likely to develop warts. These can occur almost anywhere on the body, but are relatively common on the head. Such warts are fleshy, and can grow quite rapidly. They attach to the skin by a distinct stalk and may bleed quite badly if damaged during the grooming process, for example, so you need to be aware of their presence, especially in a long-coated dog.

Swellings evident under the skin often turn out to be cysts, which can also be removed by surgery. Your vet will be able to check that such

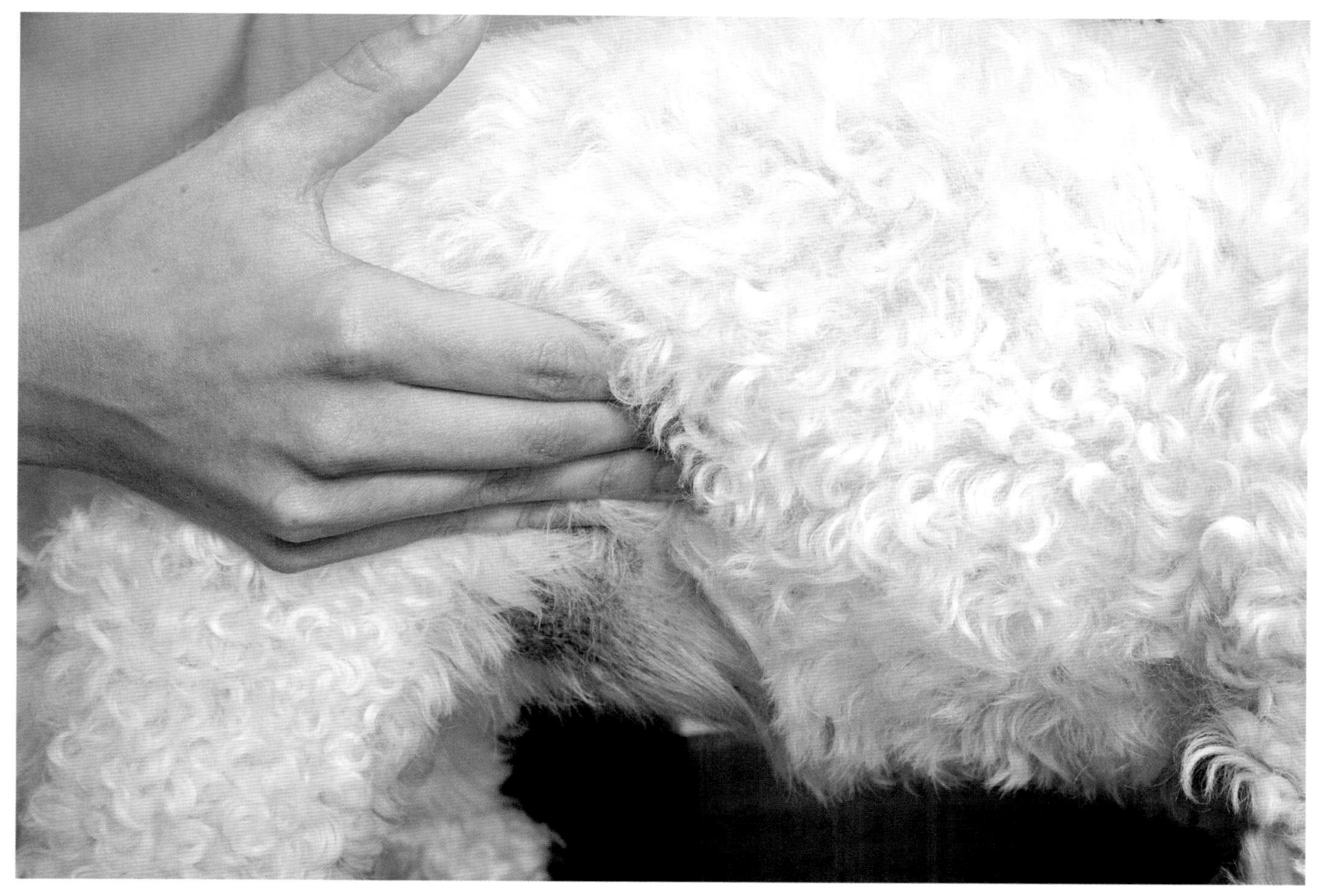

Checking the skin for possible tumours.

lumps are nothing more sinister by possibly taking a biopsy of the abnormal tissue, and sending this off to a laboratory for examination. By looking at sections under a microscope, it is often possible to tell whether a growth is benign or malignant (cancerous).

There are a number of options available to vets when treating malignancies in dogs. The skin is the most commonly affected area (four out of every ten cases), while the mammary glands are a particularly common site in the case of bitches. Surgery is an obvious option, although this is not always recommended, because of the risk of infection or the proximity of the blood supply.

The anal area, which is a relatively common site for tumours in older male dogs, is frequently dealt with by means of a technique known as cryosurgery. This entails liquid nitrogen being

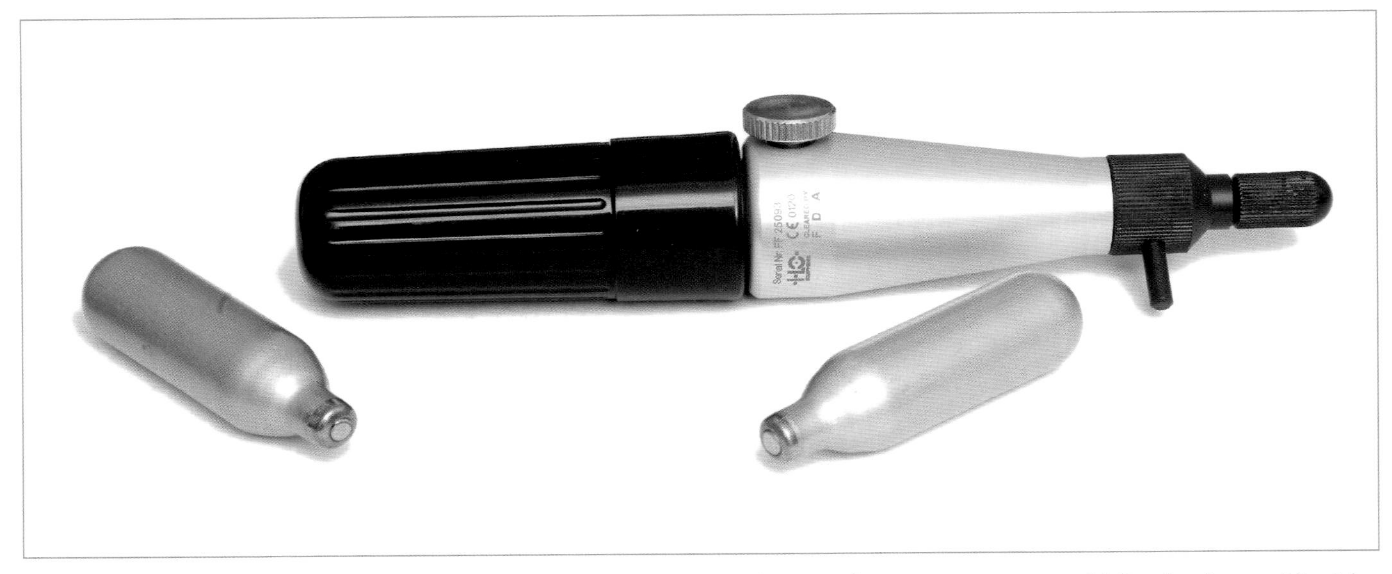

Cryosurgery equipment. Cryosurgery is a minimally invasive technique that uses extreme cold (in the form of liquid nitrogen) to destroy abnormal or diseased tissue.

applied to the site of the tumour by means of a probe, with the healthy surrounding skin masked off. The nitrogen freezes the tumour so that it eventually sloughs off, to be replaced by healthy tissue. This is a very safe option, although on occasions, more than one treatment may be required. A further advantage of cryosurgery is that there is no risk of cancerous cells entering the bloodstream, as can occur during surgery, and being carried elsewhere in the body to establish a secondary tumour or metastasis.

Drug therapy is also being used increasingly to treat certain types of tumours in dogs. These do not cause the hair loss associated with this treatment in humans. Radiotherapy is another possibility in areas where the necessary facilities are available, although you may need to be prepared to travel with your pet to a specialist treatment centre, often based at a veterinary college, for this purpose. The cost of a course of radiotherapy may well be covered by a veterinary insurance plan.

A number of factors will be involved in deciding upon the most appropriate treatment in a particular case, and even whether this should be undertaken. Discuss the options with your vet, and if you are still concerned, you can ask for a second opinion. This often applies in the case of large dogs, such as Great Danes, which have the greatest risk of cancer affecting the long bones. Amputation of the affected limb is often the most appropriate treatment, but it is disfiguring, and there is the worry about how the dog will react to her handicap after surgery. In most cases, however, there is little to fear, because dogs generally adapt surprisingly well. The strength of the other leg soon increases, compensating for the loss of the affected limb. Your vet may be willing to put you in touch with other owners whose pets have undergone similar surgery, so you can reassure yourself on the practical concerns before surgery is carried out.

If your dog has been neutered, this will eliminate the risk of malignancies affecting the testes. In the

Even a disability such as a lost limb doesn't mean your senior dog can't still enjoy an active life.

case of a bitch, spaying early on in life reduces the likelihood of mammary tumours. There is also a clear breed susceptibility to certain types of tumour, and, overall, the highest incidence is likely to be in Boxers. It has been estimated that they are, on average, about four times more susceptible to tumours in general than other breeds. Keeping a close watch on your dog, especially when grooming her, should help to detect tumours at an early stage, increasing the likelihood of successful treatment.

Urinary system

Kidney function declines in the senior dog. The kidneys play a vital role in removing the waste products of body metabolism in a solution from the blood. They also help to control the calcium : phosphorus level in the body, keeping the skeleton healthy by producing a hormone that regulates calcium absorption from the intestinal tract. Anaemia may also result in severe cases, because another hormone originating from the kidneys stimulates the production of red blood cells in the bone marrow of the body.

A number of diseases to which dogs are susceptible, such as leptospirosis, can damage the kidney tubules, as may poisons, and these will speed the normal degenerative process. The kidneys become infiltrated by fibrous tissue, giving rise to the condition described as chronic interstitial nephritis. Once about 70 per cent of the kidney tissue is affected, clinical signs of illness will become apparent. There is no way of reversing this process, but careful management of the dog's diet will help offset the worst effects.

The most obvious symptom of chronic interstitial nephritis is likely to be an increase in the dog's fluid consumption. Particularly as your dog gets older, try to keep a note of how much water she drinks each day. If you fill the bowl with a set volume of fresh water every day, you can then note how much is left at the same time on the following day. There is likely to be some variance in your dog's water consumption, however, since she is likely to drink more when fed on dry, rather than canned, food. Similarly, dogs will consume more water when the weather is hot. Even taking these factors into account, it may not always be possible to determine exactly how much water your dog is drinking, because she may choose to drink from a garden pond, or other sources of water, such as puddles, when out for a walk.

A dog's failing kidneys are unable to concentrate the urine in the way that they did when she was healthy, and so there is a correspondingly higher volume of urine, linked with the increase in water consumption. Apart from urine tests, blood sampling may be recommended to test for the level of urea in the circulation. A raised blood urea figure indicates loss of kidney function, and can be compared with the urine urea figure. Another similar comparative test can be carried out to detect the presence of an enzyme called creatinine, which accumulates in the blood as a result of kidney failure.

In relatively mild cases, dietary modifications alone may be able to stabilise your dog's condition, at least for a time. A switch to a special high protein food will lessen the build-up of waste products in the circulatory system, reducing the burden on the kidneys. The nutritive value of this protein will also serve to stem the inevitable loss of weight and condition that becomes apparent in the latter stages of chronic renal failure. At this point, your vet may prescribe the use of anabolic steroids to maintain body condition, and stimulate your dog's appetite, which is likely to start decreasing because of the build-up of urea in the circulation. Apart from changes to the protein in your dog's diet, the level of water-soluble vitamins – members of the Vitamin B group, plus Vitamin C – in the diet will need to be increased, because they are not stored in the body to any appreciable extent, and so will be flushed out more readily, due to increased urinary output. A shortage of nicotinic acid, one of the Vitamin B group, may turn the dog's tongue black, notably at the tip, and this is a symptom of long-standing renal failure.

The damage to kidney tissue will also have implications for the body's calcium levels, as mentioned previously. Skeletal weakness, sometimes described as osteodystrophy, follows, and there is likely to be a corresponding increase in the level of phosphorus in the blood. Foods produced for dogs suffering from chronic renal failure also contain low levels of phosphorus for this reason.

The general decline in health as a result of this condition means that a dog may also suffer increasingly from urinary tract infections in later life, particularly in the case of bitches, because their urethra, connecting the bladder (where urine is stored) to the outside world, is relatively short, compared with male dogs. Harmful microbes from the lower part of the urinary tract can spread to here, resulting in the inflammation described as cystitis. Typical symptoms include pain on urination, with the dog being unwilling to stay in one place while urinating for this reason. In severe cases, the urine itself may sometimes be discoloured, due to the presence of blood. Investigation of the cause will require a urine sample (see also the chapter entitled *Using your vet*). Treatment with antibiotics can also help, although recurrences are not uncommon. As with chronic renal failure, it is vital with a urinary tract infection not to restrict your dog's fluid intake in any way.

Pyometra – infection of the womb

If your bitch is not neutered, she will also become more vulnerable to infection of the womb, known as pyometra. Increased thirst can be a common sign of this condition, with infection usually arising about six weeks after the bitch's last season. Loss of appetite, and sometimes – but not always – a discharge from her vulva, are typical signs. This is a serious condition needing rapid veterinary treatment. Antibiotics and fluid therapy may be required before it is safe to operate and spay her.

Heart disease

Although heart disease is relatively common in older dogs, they do not suffer from narrowing of the coronary arteries supplying the heart muscle that is often responsible for sudden death in people. Instead, dogs tend to succumb to disease of the valves within the heart, resulting in chronic heart failure. Although the early signs may not be immediately recognisable as such, a dog may develop a persistent cough, especially after exercise. Tiredness may also be a symptom, as is a slightly swollen abdomen, caused by the pumping action of the heart being impaired.

Your vet will probably be able to detect weakened valves by listening to your dog's heart with a stethoscope; an X-ray may also be recommended to assess any corresponding enlargement of the heart. The muscle mass may increase in size, as it works harder in an attempt to correct the failure. One of the great advantages of having an older dog checked by the vet every four to six months is that such ailments should then be detected at a relatively early stage, facilitating treatment. Some breeds, particularly smaller ones such as the Cavalier King Charles Spaniel, are more prone than others to heart valve problems, whereas large dogs, like Irish Wolfhounds, may succumb to cardiomyopathy (heart muscle disease). The accompanying degeneration of the

Cavalier King Charles Spaniels are among the breeds particularly susceptible to heart valve problems.

heart muscle, in turn, leads to an abnormal rhythm. An electrocardiogram (ECG) can be very useful for investigating this type of problem, although relatively little in the way of treatment can be done in such instances.

In cases of valvular disease, however, tablets called diuretics are normally prescribed to remove excess water from the body, and so reduce the pressure on the failing heart. These are often combined with a group of drugs called cardiac glycosides, such as digitalis, that help to ensure the heart continues to contract properly. Given this type of treatment, dogs can often enjoy a number of years of reasonably good health, and lead relatively active lives during this period. Some dietary changes may assist as well, such as those designed to slim a dog, if she is overweight.

Digestive disorders

You may find that your dog suffers more frequently from digestive upsets as she grows older. These need to be investigated, since they could indicate a problem such as kidney failure, for example. If the digestive system is not functioning as well as before, your dog may suffer more from flatulence, which may be corrected by special herbal tablets produced for this purpose, or adding charcoal biscuits to her diet. Constipation may be relieved by adding a teaspoonful (5ml) of liquid paraffin to your dog's food, and, hopefully, prevented in the future by adding fibre such as a little wheat bran (sold in pet shops) to her diet. If your dog suffers repeatedly from constipation in spite of these measures, you should consult your vet in case there is an internal obstruction, such as a tumour, that is causing the problem.

Hormonal disorders

Hormones act as the body's chemical messengers, triggering specific activities in the body's organs. Several types of hormonal disorders are common in older dogs, with a number showing a distinct breed susceptibility. Diabetes mellitus (or sugar diabetes), for example, is most common in Dachshunds and Scottish Terriers. Obesity can also predispose to the development of this condition, which results from abnormalities involving the hormone called insulin, which is produced from the pancreas, a gland close to the small intestine. Insulin helps to regulate carbohydrate metabolism within the body, and in cases of diabetes mellitus, where insulin is either deficient or cannot fulfil its role adequately, the breakdown of fats is increased. This results in the accumulation of chemicals called ketone bodies in the blood, which also taint the breath, creating a sickly, sweet odour. The dog's thirst dramatically increases and, despite having a keen appetite, she loses condition, because she cannot utilise her food properly. She may also urinate more, and one of the first signs of this condition is that she is no longer able to wait until morning to do so.

Diagnosis of diabetes mellitus is quite straightforward, and is based on blood and urine tests. Treatment is possible, although you must be prepared to give your dog regular daily injections; your vet will show you how to do this. Close monitoring of her condition will be necessary, especially in the early stages after diagnosis, to ensure that the insulin dose is correct. This is usually done with test sticks that detect the presence of glucose in the urine. A regular routine will also be essential to stabilise the condition of a dog suffering from this illness, and a balanced diet – given at the same time every day – is important to prevent glucose levels rising unexpectedly. Preventing your pet from taking excessive exercise is also important, because under these circumstances, her blood glucose levels will decline dramatically.

Another form of diabetes also occurs, although it has no relation to diabetes mellitus. Known as diabetes insipidus (or water diabetes), it results from a lack of antidiuretic hormone (ADH), produced by the pituitary gland in the brain. This hormone acts on the kidneys and serves to concentrate urinary output. Diagnosis of this condition relies upon the so-called water deprivation test, during which, under veterinary supervision, water is withheld from the dog for a period of time. This would

The Dachshund is a breed that often suffers from diabetes mellitus. Sometimes, this condition can be linked to family bloodlines.

normally result in the dog's urine becoming more concentrated, thanks to the influence of ADH, but in cases of diabetes insipidus, the urine remains dilute. This type of diabetes is much rarer than sugar diabetes. Unfortunately, treatment can be costly, especially as it will be needed on an on-going basis. However, provided you can cope with the high urinary output, and ensure that your dog has sufficient water available at all times, treatment is not necessarily essential.

The thyroid glands have widespread effects on the body's metabolism. In older dogs, the hormonal output of these glands, which are located in the neck, close to the windpipe, may decline. This will cause a range of symptoms, which are likely to include unexpected weight gain, increased susceptibility to cold, and lethargy. Blood tests will confirm whether or not the dog is suffering from hypothyroidism, so that appropriate treatment, in the form of tablets, may be given, which should allow a fairly rapid improvement in your dog's condition.

The adrenal glands are located close to the kidneys. Malfunctioning of the outer layer, called the cortex, can result in an excessive output of corticosteroids. The resulting illness, called Cushing's disease, may strike in middle age. Again, it is more common in certain breeds than others, with Boxers, Poodles, and various types of terrier being most at risk. The output of corticosteroids is under the control of a hormone known as adrenocortical stimulating hormone (ACTH), which is produced from the pituitary gland at the base of the brain. A tumour here – that may not immediately be apparent in terms of the dog's behaviour – can trigger an increase in ACTH output, but the effects of the resulting upsurge in corticosteroids will soon be obvious. Typical symptoms are thinning of the coat on both sides of the body, leading to bald areas, and a pot-bellied appearance caused by an increase in the size of the liver, and the pressure of the abdominal organs, as the skin thins over this part of the body.

Testing for Cushing's disease entails administering ACTH, measuring the cortisone level in the blood, and comparing this with the resting value taken beforehand. A tumour of the pituitary or, less commonly, a tumour of the adrenal gland itself, is difficult to treat successfully. Therapy is usually directed at lowering the output of corticosteroids by means of a cytotoxic drug that destroys much of the adrenal cortex, so that although the output of ACTH may be high, only a relatively small response will follow from the adrenal glands.

Neurological disorders

In the latter stages of a tumour affecting the pituitary gland, a dog may collapse and suffer convulsions. This is very upsetting to see, because the dog often paddles helplessly on her side, in a highly distressed condition. There's not a great deal that you can do (and be aware of the risk of being bitten). You can provide something for your dog to lie on, if you can gently ease her onto this, because she may suffer loss of bladder and bowel control. Call your vet straight away for advice on when you shoud take your dog to the surgery. Veterinary help will be needed to obtain an accurate diagnosis of the cause of the problem, which could be related to the distemper virus, giving rise to what is sometimes termed 'old dog encephalitis.' Diseases affecting other body organs, such as the kidneys, may also affect the brain if they become severe. Electroencephalograms (EEGs) may be recommended to examine the pattern of electrical activity in the brain, and determine if epilepsy is the cause, although the sudden onset of epilepsy in old dogs is rare. In order to confirm the presence of a tumour, you may need to take your dog to a veterinary hospital for a brain scan as this will offer a far better way of diagnosing such problems than a traditional X-ray.

Paralysis can also strike older dogs in particular, with the hind limbs of bigger dogs often being most vulnerable. This may be the result of an intervertebral disc impinging on the spinal cord, and can often be very sudden in onset, but may respond well to rest and painkillers. Other cases

When an old dog becomes arthritic, she may need some additional support to help her get about comfortably.

may be due to bone damage, often resulting from the effects of arthritis. The dog first starts to show weakness of the limbs, which then progresses to paralysis.

In German Shepherds, there is a specific, well-recognised neurological disorder that causes degeneration of the spinal nerves, resulting in progressive weakness, paralysis and incontinence. Known as chronic degenerative radiculomyopathy (CDRM), sadly, there is nothing that can be done in the latter stages of this disease. In other cases where only the limbs are affected, it is possible to obtain a specially-designed cart that attaches to the dog's body with a harness, and provides support and mobility for her paralysed limbs. Although this may sound rather bizarre and can often draw ill-informed comments when you are out exercising your dog, there is no doubt that these carts can give dogs a new lease of life following paralysis of their hind limbs. Dogs generally adapt well to their newfound mobility, and these carts are available in several sizes to suit different breeds.

SPECIAL CARE

Although it's not possible to reverse the effects of ageing, there are an increasing number of options that will help to ensure your dog stays fit and healthy for as long as possible. It will help if you are alert to changes in your dog's overall health and level of activity, since early recognition of a problem means that it should be easier to deal with and, hopefully, remedy.

Parasites

Dogs can be afflicted by a variety of parasites, including ticks and mites, although fleas are perhaps the most common of these pests, and unfortunately also one of the easiest to acquire. Fleas are a real nuisance to both dog and owner, and older dogs can be especially susceptible to playing host to these small parasites. This is one of the reasons why it is important to groom your senior on a regular basis, since effective brushing, and careful checking down at skin level, will help you to identify the presence of fleas before a major infestation occurs. These common parasites are capable of multiplying very rapidly, with a single female flea able to produce as many as 500 eggs daily. Once these start hatching, after about ten days, you may soon be facing a plague of these insects; the situation can even escalate to the point that a pest control company may need to treat your entire house!

Older dogs can be more vulnerable to the effects of fleas because their immune system is usually less resilient than that of a younger animal. Furthermore, because your older dog may be less active, fleas have more of a chance to multiply rapidly. The risk of severe problems with fleas is highest in the summer, but these parasites can be a year-round plague due to central heating. The tiny flea larvae that hatch from the eggs seek out dark areas, often close to the dog's bed, where they feed on skin debris. Vacuuming thoroughly in this part of the home should help to remove many of them before they can pupate and change into adult fleas. This stage in the life cycle takes about three days, and the emerging fleas will then leap on to any living object, which may be your dog, or even your leg! Although fleas will not live permanently on the human body, they can inflict painful bites. Cats sharing the house with a dog are equally at risk, and can pass fleas to each other. It is therefore vital to treat a cat at the same time as a dog – although not all products are safe to use on cats, especially young kittens, so check any treatment carefully before you use it.

Fleas will cause a dog to scratch and nibble at his skin in an attempt to lessen the irritation caused by the bites of these parasites. But the situation can become far worse, because repeated exposure to the fleas' saliva sometimes results in an allergic reaction. As your dog ages, the risk of this problem is likely to increase, underlying the condition sometimes described as flea bite eczema. It will then take just a single bite to provoke a very intense reaction, often causing inflammation over quite a large area of skin. The inner area of the thighs and the base of the tail are likely to be the most vulnerable parts. Your dog will suffer great discomfort, and his persistent self-mutilation and licking will result in bald patches developing, revealing moist, reddened areas of skin beneath, which can become infected in some cases.

Effective control of fleas is vitally important, therefore, and if symptoms of flea bite allergy start to develop, rapid treatment for this condition from your vet is essential in order to desensitise the dog. Otherwise, in chronic cases, the loss of hair becomes permanent, and the skin becomes abnormally thickened. Recurrences then become more frequent, especially during the summer months when fleas are most numerous.

Another less visible danger associated with fleas is their ability to transmit other parasites, notably tapeworms. The tapeworm in the dog's gut releases eggs that adhere to the anal area, and to bedding. If a flea larva ingests one of these microscopic eggs, it will develop into an immature tapeworm within the flea. Subsequently, should the dog swallow the flea when nibbling at his skin because of the irritation here, the tapeworm will complete its development in the

Dealing with fleas

You are more likely to see flea dirt in the coat, rather than actual insects (see page 00). Nevertheless, brushing the coat thoroughly with a special flea comb will help dislodge any fleas present. Do this outside, so that any fleas you do dislodge are no longer in your house. It can be difficult to catch fleas, even when using a flea comb. Once disturbed, those that don't jump off the dog will disappear elsewhere on his body. If you do manage to catch any, drop them into a pot of water to kill them. It can help to stand your dog on white paper at this stage, too, since this will show up the dark-bodied fleas more easily when they leap off his body.

Because of the difficulty of catching fleas, a number of new approaches to controlling them have emerged in recent years. For example, special battery-operated electronic flea combs are now available. These units emit a small charge, which stuns any flea if the tooth of the comb comes into contact with the parasite, and yet will not bother or harm your dog. In fact, dogs appear to like being groomed in this way, benefiting from the massaging movements of the teeth. The motor generating the charge cuts out once a flea is hit, alerting you to the fact and enabling you to brush it out of the coat. This type of approach is particularly valuable for dogs known to suffer allergic reactions to chemical methods of flea control, which are most likely to become apparent later in life following repeated treatments.

Another approach to controlling flea numbers has been to develop products based on insect growth regulators, or 'IGRs.' These prevent female fleas from breeding successfully, or alternatively, may ensure that their eggs will not hatch. Treatments of this type are most commonly administered directly via the dog's skin. Flea control based on IGRs will not rapidly curb an explosion in flea numbers in the way that other chemicals such as sprays should do, but they are very effective at preventing a significant flea population developing in the longer term. Other methods of control include various types of flea trap that are designed to lure fleas into them, trapping the insects on sticky paper. There are also sprays and powders, some of which are applied directly to the coat, and collars that release a chemical to eradicate fleas. An insecticidal canine shampoo is also available. Check with your vet before using any flea treatments, however.

You should also wash your pet's bedding, including the bed itself, when you treat your dog. As an additional precaution, you may also want to clean the carpet, but check beforehand that any chemicals used for this purpose will not affect its colour.

When using a flea shampoo, it's usually best to start by lathering the neck thoroughly. This will create a barrier, preventing fleas from migrating to the ears and head, which are more difficult to treat with shampoo. Then wet and lather the rest of the body with shampoo. Always follow the instructions closely; wearing plastic gloves may be advisable.

dog's intestinal tract. Without fleas to act as an intermediate host, however, the common tapeworm (Dipylidum caninum) represents no hazard to dogs. If your dog has suffered from fleas, however, it is also a good idea to treat him for tapeworms as well. Tablets for this purpose can be obtained from your vet or pet shop.

Ticks can also be a problem, especially in the summer months. See the chapter entitled *General care* for how to safely remove these from your dog's skin. Skin mites are not normally a problem in older dogs. However, dogs can sometimes pick up 'harvest mites' when out walking, particularly in the countryside. These are actually the larval stage of the trombiculid mites' life cycle. The tiny larvae affect the feet, and can cause a very intense, painful irritation between the toes. Treatment from your vet is the best way of dealing with these parasites. Ear mites are more common, particularly if your dog has suffered from ear problems previously, and will make your dog scratch continually in this area.

These parasites can often be linked with bacterial and fungal infections within the ear canals: again, your vet will need to examine your dog to prescribe the most appropriate treatment.

A dog that has collapsed from heat stroke is covered in a special cooling blanket that will help to lower his body temperature.

Watching the weather

During the summer, it's a good idea to avoid exercising your pet around the hottest part of the day, especially as he grows older and becomes less active. This applies particularly if he is suffering from a heart complaint, because the effects of heat stroke on top of this may prove fatal. Breeds with short noses, such as Bulldogs, are especially at risk of overheating in hot weather, simply because they are less able to cool themselves, other than by panting. The relatively short length of their nasal passages means that the amount of moisture evaporating from here is reduced, compared with other dogs, and so their ability to control their body temperature is diminished.

During the colder months of the year, it's a good idea to obtain a coat for your dog, which should help him stay relatively warm and dry, irrespective of the weather, and also ease painful joints. Certain breeds, such as the Whippet, are especially vulnerable to becoming chilled, simply because of their lack of a dense undercoat. A wide variety of dog coats are normally available from pet shops and similar outlets, with some being more weather-resistant than others. In order to ensure a good fit, try to take your dog along with you when you make the purchase. Failing this, measure your pet from the back of his neck to the base of the tail, and also around his chest, so you can select a coat that should fit comfortably. If you often need to exercise him after dark, have bright, reflective, fluorescent strips on the coat which will alert others/traffic to your presence.

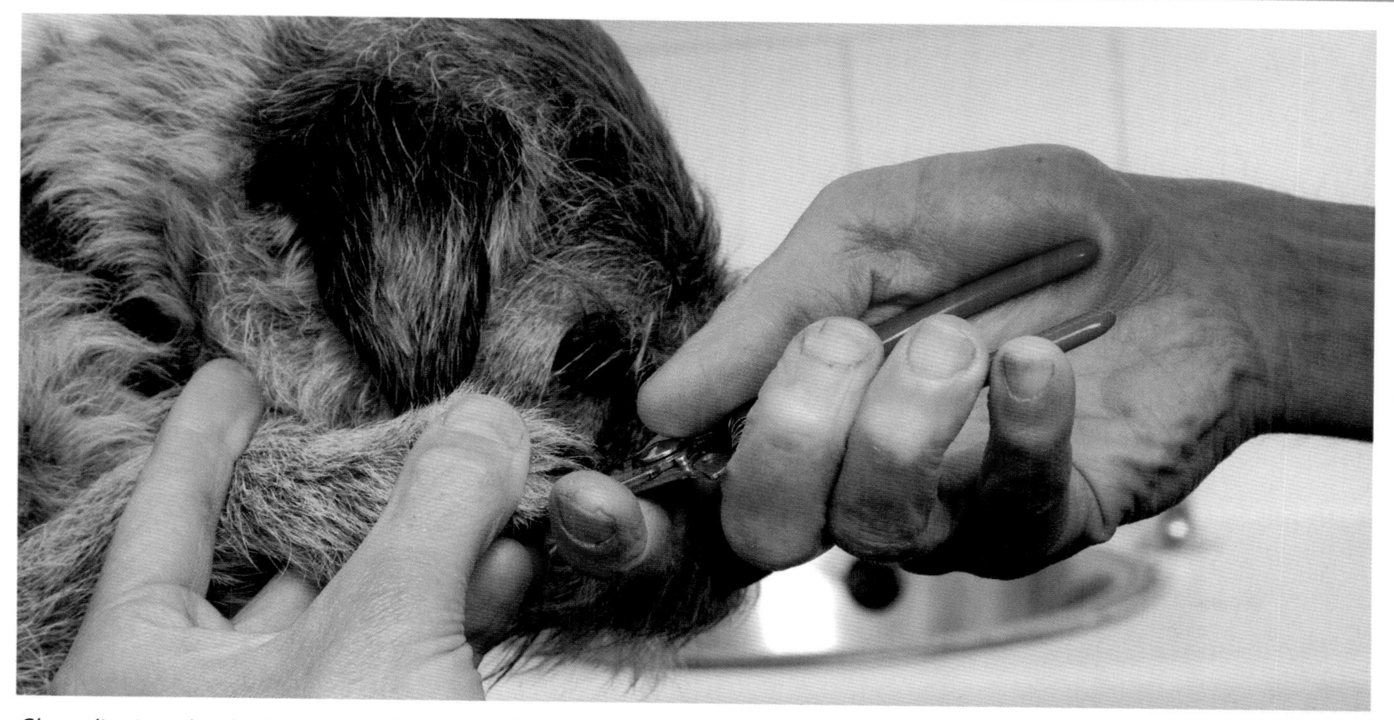

Claw clipping should form part of your dog's regular grooming routine. The claws of older dogs are more likely to become overgrown, since they are possibly not being worn down by as much walking. Filing may be an option also.

Grooming

Dogs have thicker coats over the winter period, which they will moult during the spring as the weather warms up. Older dogs may be less able to shed this hair if they are no longer very active, and so grooming assumes greater significance for them. The time required, and the way in which the fur is groomed, will depend very much on your dog's coat. A daily grooming with a brush and comb will help prevent the fur of long-coated breeds from becoming matted. A dense undercoat can be stripped out with a wire brush. For short-coated breeds, a hound glove can help to give the coat a good gloss. Seek advice from a professional groomer if you want to know the best way to groom your dog.

Some older male dogs with hip weaknesses may need to be bathed more frequently than usual, simply because they can sometimes have difficulty lifting one of their hind legs to urinate successfully. As a consequence, urine may stain the surrounding area of the coat. Similarly, older dogs of both sexes may encounter more difficulty than usual when defecating, and this may also mean that additional care is required, particularly in the case of long-coated animals. Any staining of the fur is likely to attract flies that may lay their eggs in this part of the coat which, if not spotted at this stage, will soon hatch into maggots that will bore into the dog's flesh and release harmful toxins. Rapid veterinary treatment will be needed under these circumstances.

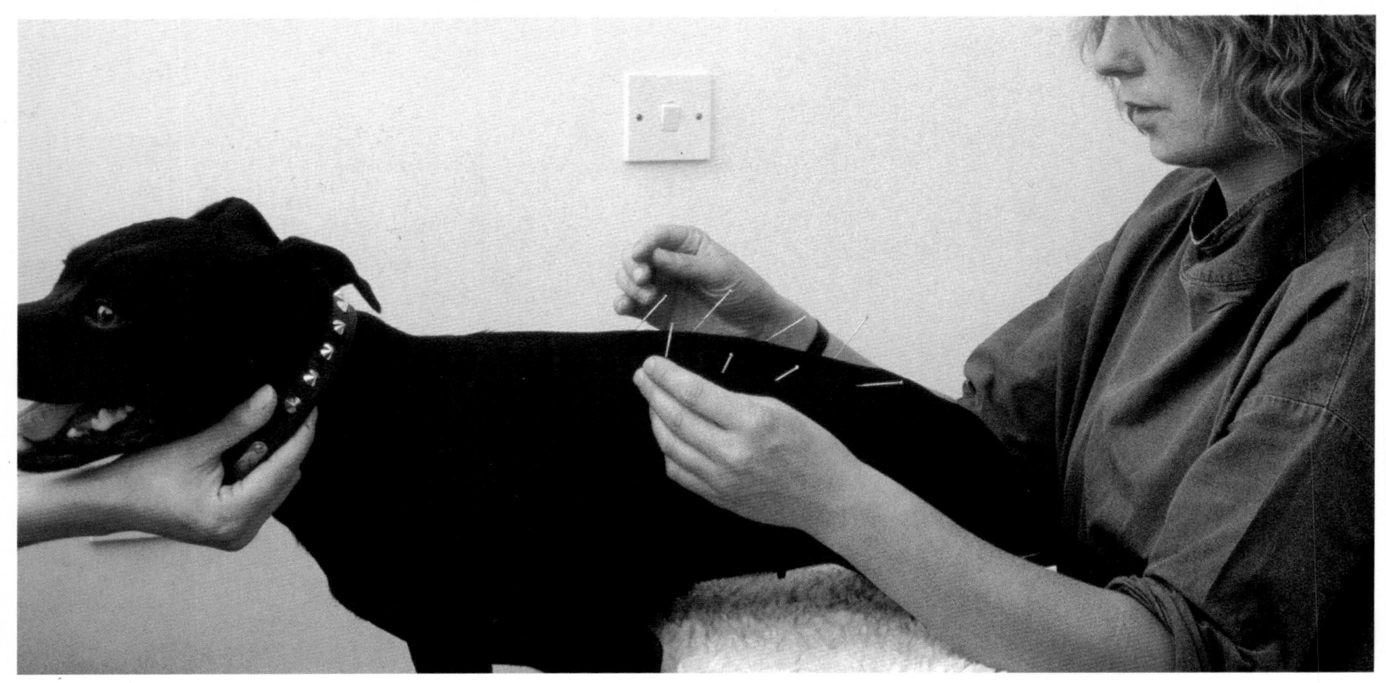

An experienced practitioner performing acupuncture on a Staffordshire Bull Terrier. The technique involves inserting needles in parts of the body and manipulating them in order to relieve pain.

Mobility issues and therapy

Hip dysplasia may be less common than it was once, thanks to techniques such as screening breeding dogs for the condition and using 'hip scoring' based on X-ray images. However, arthritic changes affecting the hip joints especially are not uncommon in older dogs. There is little that can be done by way of direct treatment, but ensuring that your dog is not overweight – which will increase the burden on the joints – is significant. Painkillers used in accordance with your vet's advice may be required on occasion, and acupuncture may also be beneficial, particularly for a relatively calm dog. (See also *My dog has hip dysplasia ... but lives life to the full!* published by Hubble and Hattie.)

A range of dietary supplements can be given to ease joint pain, and these are available as liquids, powders and capsules. They contain various constituents such as chondroitin, glucosamine, MSM (a sulphur-based compound), and hyaluronic acid, and help to improve the amount of synovial fluid, which lubricates the joints and encourages the development of protective cartilage within the joints, making them less painful. Always select a brand developed for dogs, rather than for humans, since they are more likely to be palatable and are easier to administer via food, for example. Various herbal products are also available for decreasing inflammation, and these may contain plants such as yucca, comfrey, and devil's claw. If you grow plants such as comfrey, you may even find that your dog will browse on them in the garden, apparently recognising the beneficial effects.

Hip dysplasia

If your dog has a history of hip dysplasia, the effects are likely to become worse with age. In a normal hip joint (far left), the head of the femur fits snugly into the cup-shaped joint known as the acetabulum. However, when the cup is shallow (centre), abnormal wear occurs, making movement painful. Over time, arthritis may affect the joint as well (right). Be sure your dog is not overweight, thus avoiding unnecessary stress on the hips. Hydrotherapy can be a useful form of exercise for dogs suffering from this condition.

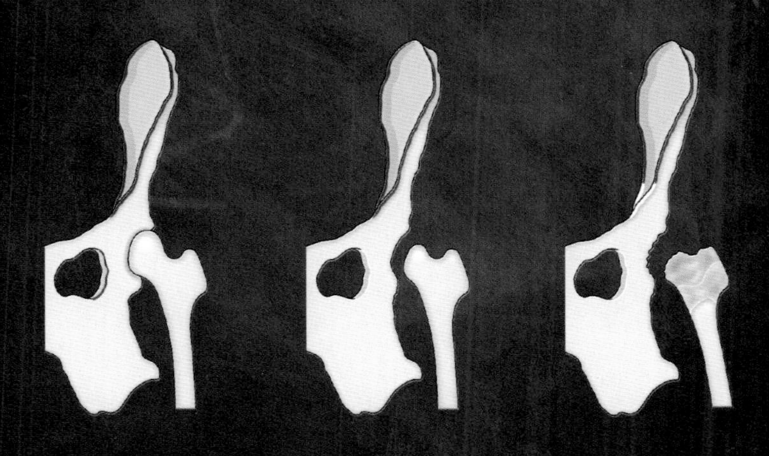

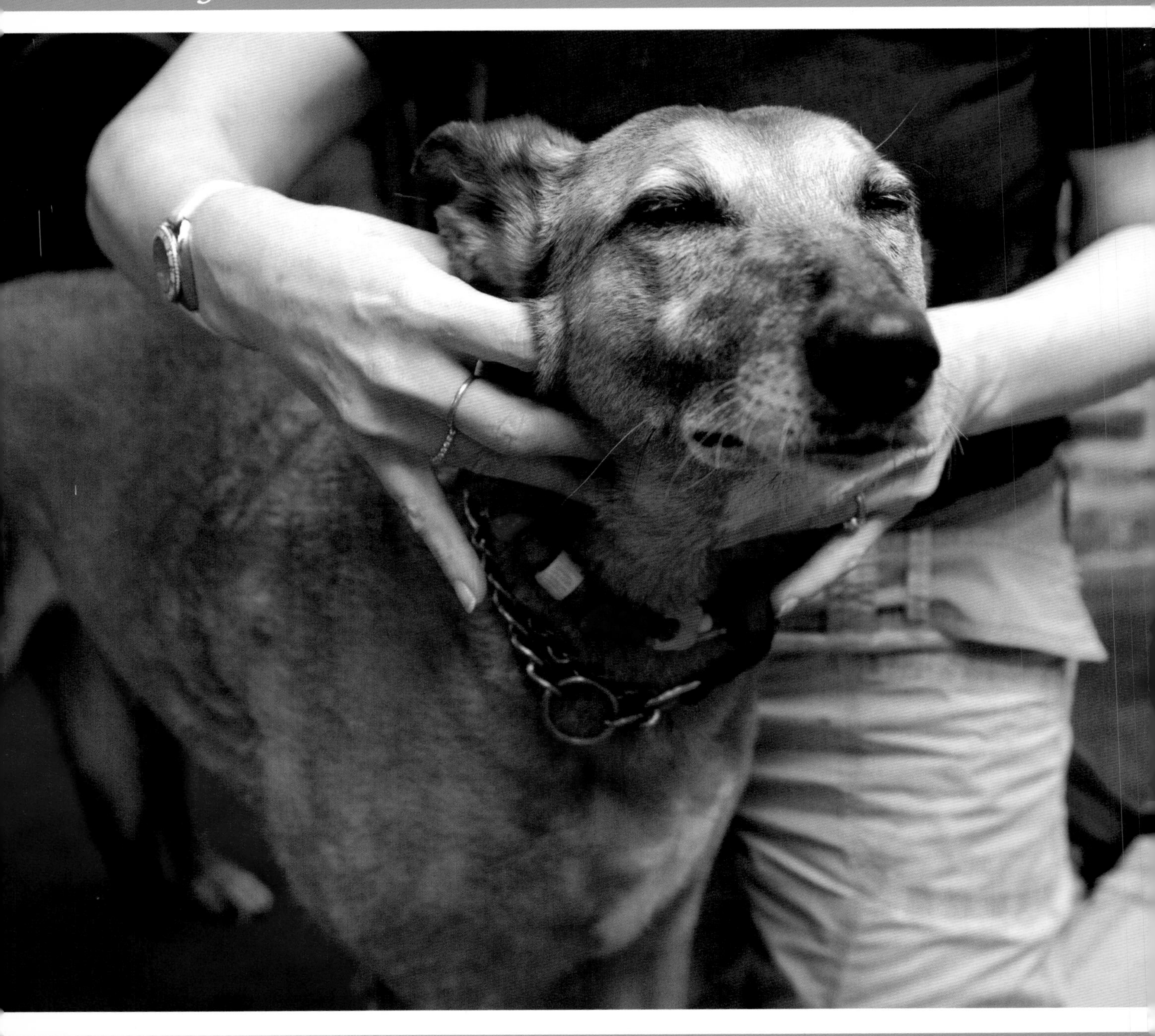

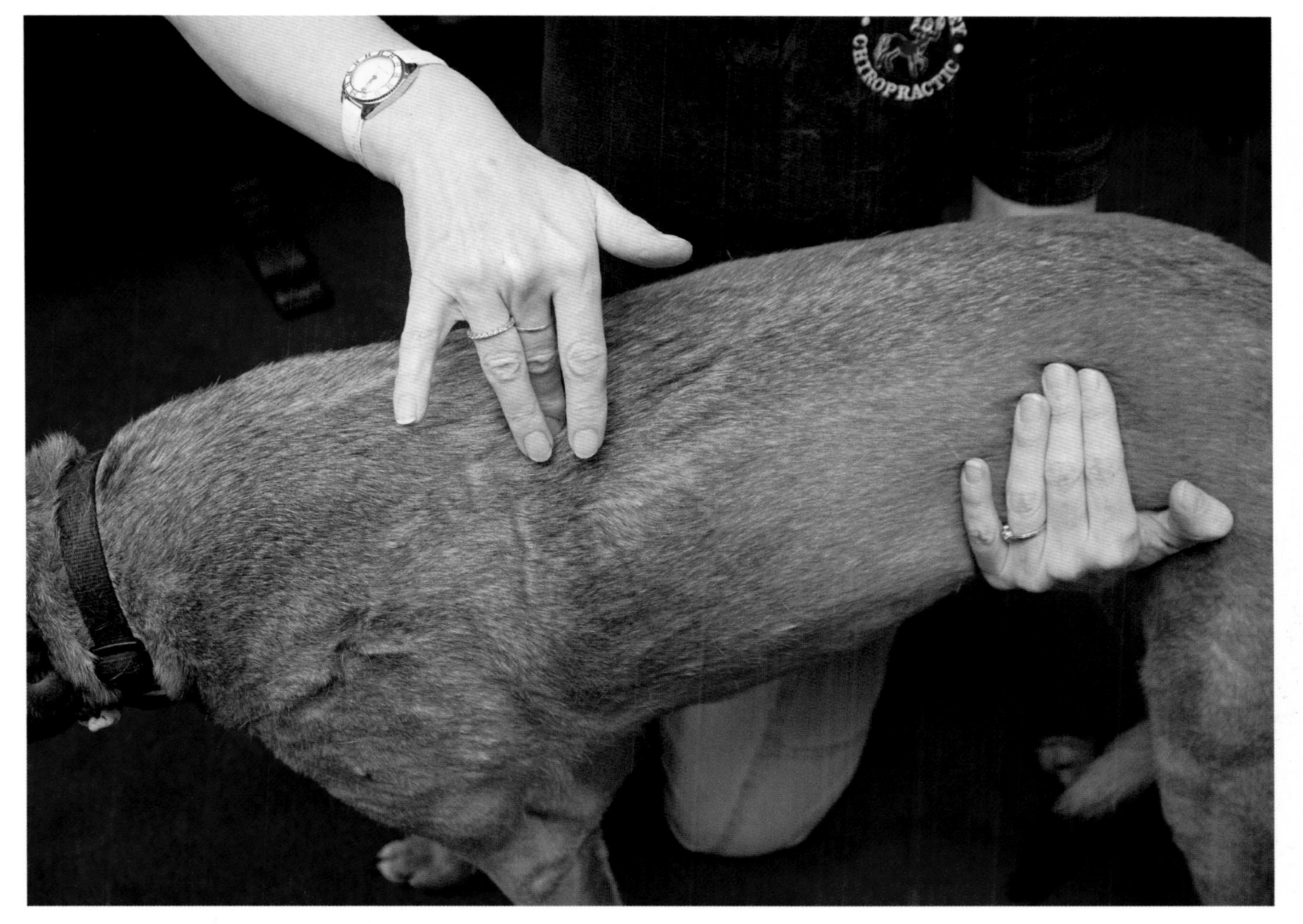

(Above and opposite) Manipulative chiropractic techniques are sometimes used to help dogs suffering from skeletal problems.

Direct intervention can be beneficial, too. Treatments such as acupuncture and chiropractic therapy are especially useful where spinal problems are involved; these tend to be most common in long-bodied breeds such as Dachshunds. Your vet will be able to refer you to a practitioner in this field. If you have a health insurance policy for your dog, you may find that cover is provided for certain complimentary therapies.

Ensuring that elderly dogs take exercise is important, even in the case of individuals afflicted by joint disorders. This has been made much easier

A properly managed hydrotherapy session can be a very beneficial form of exercise. It can also help as part of the programme of recovery following some types of injury or post-surgery. Note the lifejacket being worn here, even though dogs are natural swimmers.

Another method of hydrotherapy uses a shallow vessel called a water walker. The dog is able to perform normal walking activities while having his body weight partly supported by the water.

thanks to the growing number of hydrotherapy pools that now exist for dogs. Again, your vet should be able to direct you to one within your area. Most dogs enjoy swimming, and sessions take place in warm water. Even dogs badly afflicted by joint pain can benefit from this type of exercise, because water acts as a buoyancy aid, supporting their weight, which means that movement is far less painful for your four-legged friend. Sessions need to be booked in advance, and although at first it may be advisable to allow your dog to swim for only a short time, this period can be increased as his level of fitness improves.

Massage can also help to ease joint pain, as well as the stress associated with it, by acting as an holistic therapy. (See *The Complete Dog Massage Manual* published by Hubble and Hattie.) Several forms are recognized, with traditional massage helping to improve circulation and muscle tone. One form, reiki, relies upon chi – the body's energy – in combination with specific energy centres. This Japanese technique can bring comfort to

(Above and opposite) Massage is another important and effective therapy for treating or easing many types of musculoskeletal problems in the older dog. You can perform some of these techniques yourself at home, but it is essential that you follow the instructions in a good canine massage manual. Alternatively, you could learn the techniques from a qualified expert first, as performing massage incorrectly may do more harm than good.

dogs afflicted by a terminal illness. Acupressure is another method that concentrates on particular parts of the body, identified by practitioners as energy centres. Such methods may also be useful in aiding recuperation after illness or surgery, as well as being relaxing and providing emotional support for an ailing pet.

Lifestyle changes are also important. You may want to replace your dog's bed with a bean bag containing polystyrene granules or a similar material. Your dog can then find the position in which he is most comfortable – for instance, stretching out, rather than having to curl up.

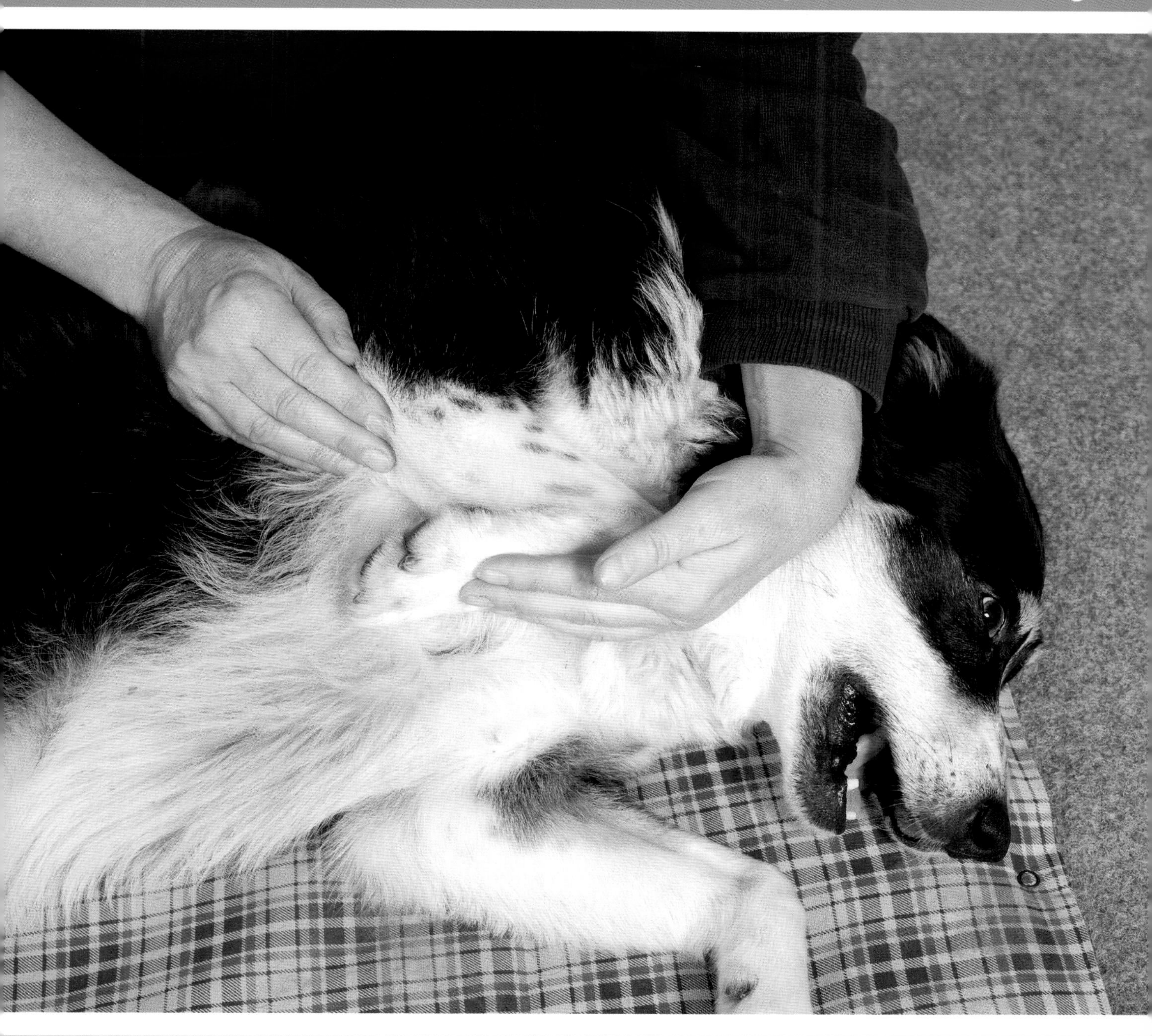

This deaf Springer has learned to respond to visual signals instead of voice commands.

Declining senses

You may notice as your dog becomes older that his hearing will decline, which can lead to disorientation, especially when out for a walk. You are likely to find that he begins to appear unresponsive around the home, too, not always hearing when you are calling. There is really very little that can be done under these circumstances other than to try to accommodate your dog's changing behaviour (but see *My dog is deaf ... but lives life to the full!*, published by Hubble and Hattie).

Older dogs tend to stray less than when they were younger, but even so, encourage your dog to stay closer to you when you are out for a walk. It is not just you that an elderly dog may not hear – approaching vehicles can also represent a greater danger. Therefore, always keep your dog on the side farthest from the road when walking along a pavement, so he cannot step out into the traffic.

Blindness is another handicap that can afflict older dogs, and again, there is little that can be done. Dogs, generally, are less reliant on sight than we are, and will prove far more adaptable under

A dog with failing eyesight or hearing should be kept as far away from roads as possible when out walking.

these circumstances because they rely to a greater extent on other senses, too, including scent and even their whiskers. These specialised sensory hairs provide them with close-up information about their surroundings. It is obviously important to get your pet's eyesight checked by your vet if you suspect that it is failing, but since sight loss tends to occur gradually, dogs usually adapt very well. Avoid any sudden changes in the home environment, such as moving furniture around in a way that blocks areas that were previously clear of obstructions, so your dog will be able to adjust to his loss of vision. The same applies to objects in the garden. Take care when out for walks, speaking repeatedly to your dog and stroking him reassuringly, so he doesn't feel isolated and disorientated. Otherwise, he is likely to start barking repeatedly, as he tries to establish your whereabouts. (See also *My dog is blind ... but lives life to the full!*, published by Hubble and Hattie.)

WHEN IT'S TIME TO SAY GOODBYE

Sadly, at some stage, all owners have to recognise the clear signs of an ageing dog.

Unless a dog dies suddenly and unexpectedly, eventually, even the fittest and most active of seniors reaches the stage when she begins to show signs that her life is nearing an end. This may be because an ongoing illness has become progressively, or even suddenly worse, or it may be due to some sudden traumatic event, such as paralysis of her legs. It is never easy taking the decision to part with your friend and companion when such a time comes, but you can always rely on the support of your vet, whose experience and kindness can be a great comfort in dealing with the situation. The most important thing is not to prolong your dog's suffering once her quality of life is seriously compromised.

Some guidelines

The likelihood is that you will know almost instinctively when you must take the decision to have your dog put to sleep, but there are significant questions that you should ask first. Does she still have an obvious interest in life, respond to her name, and is able to move around without

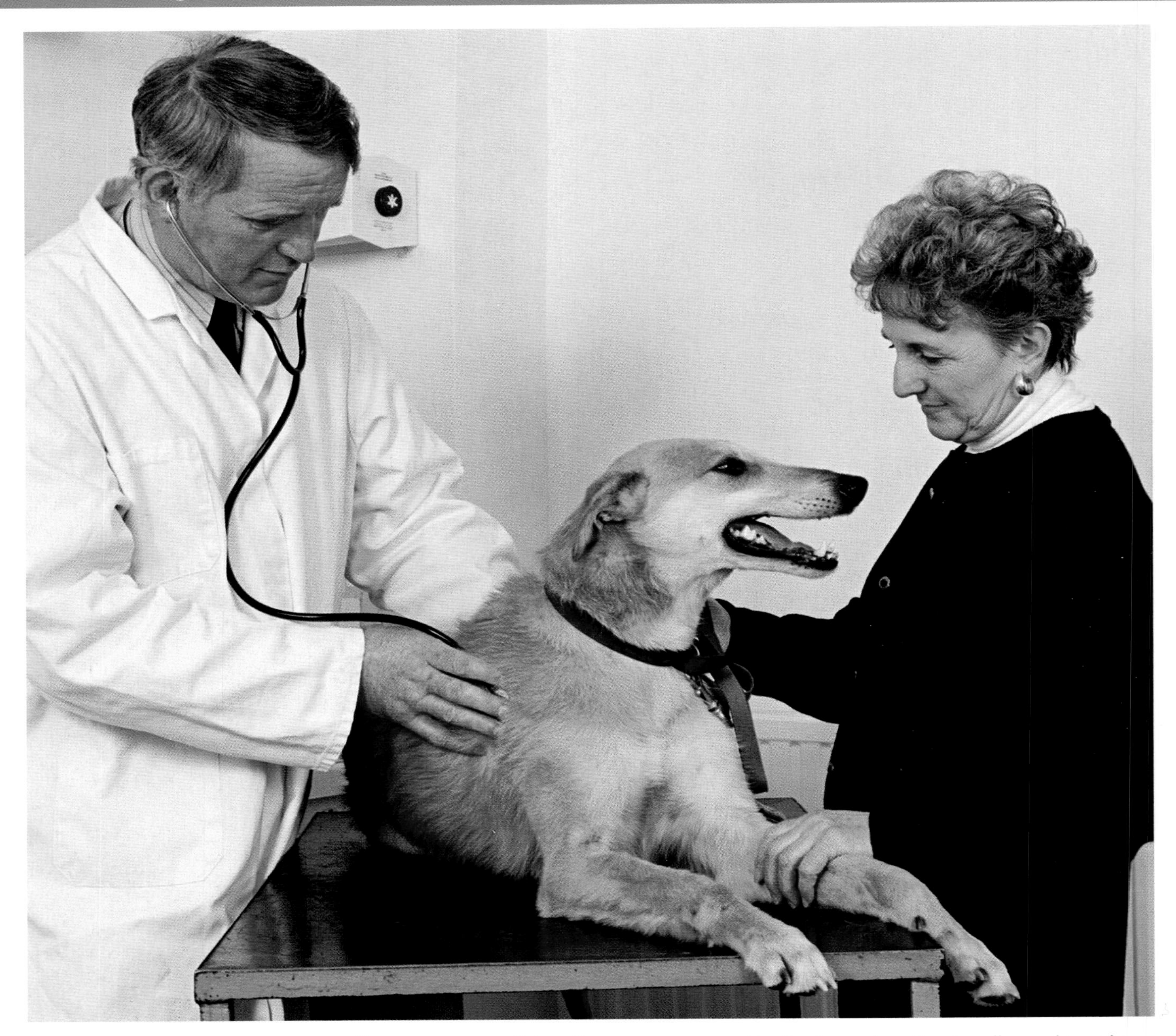

When the time comes to say goodbye, your vet will handle matters sympathetically and professionally, making the whole process much easier for you and your companion.

experiencing great difficulty? Is she in obvious pain? Is she eating and drinking normally? Has she become incontinent, or appear not to recognise you? Although you may feel able to cope with incontinence, or a dog that has difficulty in moving around, it is much harder in cases where a beloved animal no longer responds to her family. Your dog will want to stay with you as long as she can, but there may well come a time when she will have had enough and the kindest thing to do is let her go.

Don't overlook the feelings of the family, either, because seeing a dog that was previously healthy and active reduced to a distressed state can be very traumatic for children. It can also put pressure on relationships. There is a further risk that a dog who is not fully aware of her surroundings, and in pain, may become unexpectedly aggressive, posing a particular threat to children.

If your dog has been ill for a while, your vet may need to carry out exploratory surgery, since there is only so much that can be done with non-invasive tests. The surgery may reveal an inoperable condition, and generally, in this situation, it is better to arrange with your vet that your dog does not regain consciousness. Euthanasia is not a distressing process in itself, since it effectively entails giving the dog an overdose of barbiturate into a vein. If a dog is already anaesthetised, this will simply cause her to stop breathing.

Making a decision

While you should not prolong the life of a pet who is suffering, nor should you feel pressured into making that final decision before you are ready. Try to weigh up the options – although if your vet says that nothing that can be done, this will inevitably be the case. There is always an inclination on hearing this sad news to think of obtaining a second opinion, but there is usually little to be gained, other than causing your treasured companion further discomfort. Second opinion referrals are really only useful in instances where a condition has proved unresponsive to treatment, or is not straightforward to diagnose, rather than at the end of a dog's life.

Having taken the decision to have your dog put to sleep, you may be asked to sign a consent form, just as you would for an operation. This is a standard procedure in most practices. You may decide that you want your vet to visit your home in order to carry out the task, and this will certainly be the least distressing situation for your dog, and very possibly for you, too. The receptionist at the practice can arrange an appointment right at the end of a surgery, however, so you do not have to sit in a waiting room full of people.

Your vet can make the necessary arrangements for you, with regard to what happens to your dog afterwards, although you might want to organise this yourself. It is possible to take your dog to a range of companies that offer individual cremation or burial, knowing only too well how painful and upsetting this part of the process will be. If your dog is cremated, the ashes will be given to you if you wish to have them.

Coping with grief

Expect to feel upset and to grieve for your pet – this is quite natural. After all, your dog may well have been part of your daily life for more than a decade, so you're likely to feel in a state of emotional turmoil for some time. Initially, there will be a strong sense of loss and emptiness, reinforced by the fact that your daily routine has changed significantly. You may also start to feel guilty, thinking that you could have done more for your pet to prolong her life. This is almost certainly without any foundation, but it can be a very strong emotion, which may also give rise to anger or resentment as far as your vet is concerned; you may find yourself looking back at the treatment your dog received and feeling that more could have been done. Again, this is a natural reaction experienced by many grieving pet owners.

Recognising the trauma that the death of a pet can cause, many veterinary practices now offer counselling sessions. It's also possible, in some areas, to speak on the phone with a pet bereavement counsellor. Time is a great healer, and, in due course, hopefully, you will be able to look back on your dog's life and remember the good times that you shared, without being overwhelmed by sadness.

You may also feel that, in due course, you want to give a home to another dog, even if, initially, your thoughts are that no other animal could replace your lost friend. The important thing to bear in mind is that, whether you eventually obtain a puppy or an older dog, she will be an individual in her own right, rather than a replacement. There is no hard and fast rule that says how long you should wait before starting to look for another dog. Some people will find their home is unbearably empty without a canine presence, and so seek another soon afterwards, whereas other people may choose to wait several months or longer.

It's natural that you will grieve for the loss of your canine friend, but in time, hopefully, you will come to remember her for the good times when she was fit and active.

It may be difficult to consider having another dog when you have just lost a dear canine friend, but it is surprising how quickly a new puppy, or an older, needy dog, can bring a new interest and focus into our lives.

GLOSSARY OF TERMS

Words in SMALL CAPITALS refer to other glossary entries.

Agility exercise
A form of exercise for dogs that involves them running through and over obstacles, around hurdles, and so on.

Allergy
A hypersensitivity to certain substances (called allergens), often resulting in a reaction such as a skin rash on the part of the sufferer.

Anal glands
Two small pouches or sacs situated around the anal opening which are used for scent marking. Dogs can experience discomfort when the glands do not empty naturally through DEFECATION.

Arthritis
A term used to describe a number of painful conditions of the joints and bones. Osteoarthritis, the most common form, occurs when cartilage between the bones wears away, causing bone to rub on bone in the joints. In rheumatoid arthritis, the body's immune system destroys the joint, causing swelling and pain.

Blood circulation
The flow of blood around the body.

Car ramp
A device, often folding, that can be used to enable a dog to access a vehicle more easily.

Cataract
A clouding that develops in the lens of the eye, usually appearing as a whitish colour in one or both of the eyes, and progressively impairing vision.

Cervix
The neck of the UTERUS.

Cognitive dysfunction syndrome
A deterioration in brain function, similar to Alzheimer's disease in humans, that is sometimes seen in dogs.

Complete food
A specially formulated type of dog food that contains all the necessary nutrients for a healthy diet.

Cushing's disease
A condition caused by a dog's adrenal glands overproducing the hormone cortisol.

Autonomic nervous system
The non-conscious part of the PERIPHERAL NERVOUS SYSTEM, controlling functions such as heart rate, heat regulation, digestion and urination.

Central nervous system
The part of the nervous system consisting of the brain and the main nerves running from the spinal cord.

Connective tissue
A type of tissue in the body that binds and supports other internal structures.

Contraindication
A condition that increases the risk of a surgical or therapeutic procedure.

Defecation
The act of voiding faeces from the bowels.

Dilation (of blood vessel)
The widening of a blood vessel, due to relaxation of the muscular wall of the vessel.

Flea
A small, parasitic, blood-sucking insect that lives in the fur of many animals, including dogs.

Flyball
A type of relay race for dogs that involves catching

a ball and then running over hurdles, back to the starting point.

Hip dysplasia
A commonly diagnosed arthritic condition in which the leg bone does not fit properly into the hip socket.

Homeostasis
A state of equilibrium within the body's internal environment, maintained by its regulatory system, irrespective of external conditions.

Hydrotherapy
A therapy that involves immersing an animal, such as a dog, totally or partially in water to help his fitness or to aid rehabilitation following injury or surgery.

Immune system
The body's natural defence system against foreign bodies.

Inflammation
The biological reaction of body tissues to harmful stimuli, such as damaged cells, irritants or infection.

Joint
The site within the body where two bones meet and where movement can usually occur. Examples of joints include the knee, elbow and hip.

Massage
The manipulation of the soft tissues of the body.

Metabolism
The body's combined physical and chemical processes.

Mite
A small, parasitic animal that lives in the skin of many animals, including dogs, and can cause mange.

Muscle
Specialised body tissue that can contract and relax, and thus produce movement.

Nuclear sclerosis
A benign cloudy appearance of the eyes that can seem similar to CATARACTS.

Palpation
An examination by touch, carried out by a vet or other medical practitioner.

Peripheral nervous system
The part of the nervous system that innervates the body's systems and organs. See also CENTRAL NERVOUS SYSTEM.

Physiological
Referring to the functions of living systems.

Pseudopregnancy
A condition in which a female animal, such as a dog, exhibits the signs of pregnancy – including certain changes in behaviour – without actually being pregnant.

Psychological
Referring to mental functions, behaviours and perceptions.

Rheumatism
A non-specific term often used to describe medical conditions affecting the joints and associated MUSCLES and other tissues.

Sebaceous gland
A gland in the skin, usually associated with a hair follicle, that secretes oil (sebum) to lubricate the coat.

Therapy
The treatment of a disease or other ailment by some form of curative process.

Skeletal system
The rigid framework of bones within the body of animals like fish, reptiles, birds and mammals (including dogs). It consists of the appendicular skeleton (the fore and hind limbs, the shoulder and the pelvis) and the axial skeleton (the skull, the ear and throat bones, the ribs and the vertebrae).

Spasm
A sudden and involuntary contraction of a MUSCLE or a group of muscles.

Tick
A small parasite that burrows its head into the skin of many animals, such as dogs, and sucks their blood.

Urinary system
The part of the body, chiefly comprising the kidneys, that controls the water volume and chemical composition of the body's fluids.

Urination
The act of disposing of urine from the bladder.

Uterus
The sac-like structure in which the embryo develops.

Voluntary muscle
MUSCLE that is under conscious control. Also called skeletal muscle.

Further reading and information

Books

The Complete Dog Massage Manual – Gentle Dog Care by Julia Robertson
Dog Relax – relaxed dogs, relaxed owners by Sabina Pilguj
Dog Games – stimulating play to entertain your dog and you by Christiane Blenski
My dog is blind – but lives life to the full! by Nicole Horsky
My dog is deaf – but lives life to the full! by Jennifer Willms
My dog has hip dysplasia – but lives life to the full! by Kirsten Hausler & Barbara Friedrich
Swim to recovery: canine hydrotherapy healing – Gentle Dog Care by Emily Wong
All published by Hubble & Hattie

Websites

www.dignitypetcrem.co.uk
www.abbeyglen.com
www.pet-crematorium.co.uk
www.ageespetcrematorium.com
www.sleepymeadow.co.uk
www.sunnyfields.org.uk
www.caninetherapy.co.uk
www.animalmagichydrotherapy.co.uk

978-1-845843-87-8 £14.99*

978-1-845843-33-5 £14.99*

978-1-845843-45-8 £9.99*

978-1-845843-86-1 £4.99*

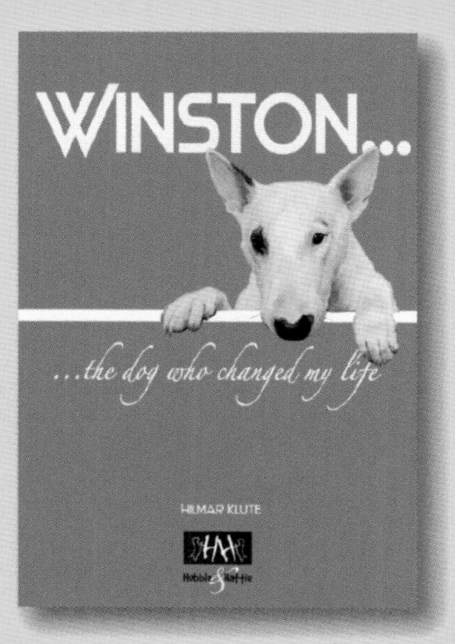

978-1-845842-74-1 £9.99*

978-1-845844-06-6 £9.99*

978-1-845842-88-8 £9.99*

978-1-845843-84-7 £9.99*

978-1-845843-85-4 £9.99*

www.hubbleandhattie.com

978-1-845840-72-3 £9.99*

978-1-845842-92-5 £12.99*

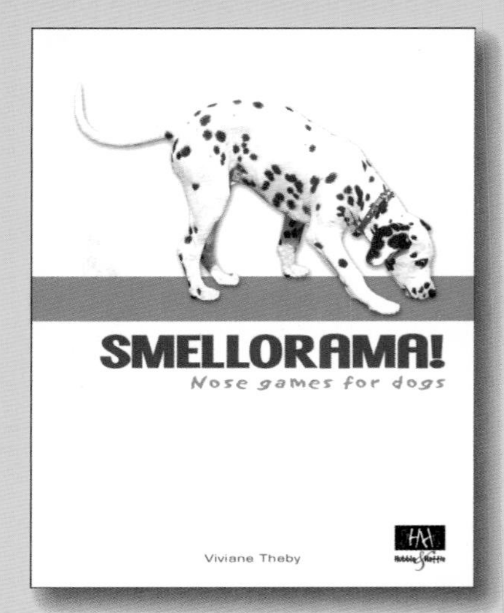

978- 1-845842-93-2 £9.99*

For more info on Hubble and Hattie books, visit our website at www.hubbleandhattie.com
email info@hubbleandhattie.com • tel 44 (0)1305 260068 • *prices subject to change • p&p extra

Index

NOTES